REVEALING WATERMARKS

HOW TO ENHANCE THE SECURITY OF HAND-MADE PAPER ITEMS AND REVEAL HIDDEN DATA

REVEALING WATERMARKS

HOW TO ENHANCE THE SECURITY OF HAND-MADE PAPER ITEMS AND REVEAL HIDDEN DATA

IAN CHRISTIE-MILLER

BOSTON
2021

Library of Congress Cataloging-in-Publication Data

Names: Christie-Miller, Ian, author.
Title: Revealing watermarks : how to enhance the security of hand-made paper items and reveal hidden data / Ian Christie-Miller.
Description: Boston : Academic Studies Press, 2021. | Includes bibliographical references.
Identifiers: LCCN 2021011433 (print) | LCCN 2021011434 (ebook) | ISBN 9781644696248 (hardback) | ISBN 9781644696255 (adobe pdf) | ISBN 9781644696262 (epub)
Subjects: LCSH: Watermarks--Europe--Identification. | Handmade paper--Europe--History--16th century. | Paper--Analysis--Data processing. | Image processing--Digital techniques.
Classification: LCC Z237 .C47 2021 (print) | LCC Z237 (ebook) | DDC 676/.28027--dc23
LC record available at https://lccn.loc.gov/2021011433
LC ebook record available at https://lccn.loc.gov/2021011434

ISBN 9781644696248 (hardback)
ISBN 9781644696255 (adobe pdf)
ISBN 9781644696262 (epub)

Book design by Kryon Publishing Services
www.kryonpublishing.com

Cover design by Ivan Grave

Published by Academic Studies Press
1577 Beacon Street
Brookline, MA 02446, USA
press@academicstudiespress.com
www.academicstudiespress.com

Contents

Acknowledgements

Grateful thanks and words of appreciation need to be recorded to:

Mika Hakkarainen Special Collection Curator Helsinki for supply of back-lit images of *ABCKiria*.

Claudia Ludwig and Lucy Finn at Sotheby's, New York for locating and supplying the 26th November 2013 *The Bay Psalm Book Catalogue*.

Dr. Brent Elliott and Ms. Debbie Lane, both formerly of the Royal Horticultural Society, who were leaders in recognizing, encouraging and supporting watermark research at the Lindley Library.

Staff at the Vilnius University Library for not only allowing the imaging of *Cathechismus,* 1547 but also enabling it. Aušrinė Aurelia Apanavičiūtė for her recordings of the music in the Vilnius copy and for the ready approval for those recordings granted by Father Petras Tverijonas at St Casimir's Lithuanian Roman Catholic Church in London.

Mr. Simon Green for his expert guidance about paper moulds.

Allard Pierson at the University of Amsterdam for efficiently supplying the Sheremetew watermark image, and to Alexander Khmelevsky for the fulsome comments on the Sheremetew family.

Dr Fred Hocker, Head of Research, Vasamuseet Stockholm for his immediate and most helpful information about cannons and heraldry.

The Bibliographical Society, The Paul Mellon Centre for Studies in British Art and The Lutheran Church—Missouri Synod for their considerable and essential funding support.

Charlotte Tancin at the Hunt Institute for Botanical Documentation, Carnegie Mellon University, to Catherine Sambrook at King's College London, Foyle Special Collections Library and to Stephen Tabor at the Huntington Library for their invaluable suggestions for improving the early draft.

Alessandra Anzani of Academic Studies Press, who has patiently guided the production of the book through many months.

Introduction

Researchers have long appreciated the potential of watermark research, but it is the advent of digital resources which is allowing that potential to be achieved. Not only is security being improved, but valuable research resources are being created.

The First Part of this book opens with a description of the *PaperPrint* method, devised by the author, as used at the Lindley Library, Royal Horticultural Society for prints, at Marsh's Library, Dublin for copies of a Cyrillic printed book, and at the Boston Public Library for *The Bay Psalm Book*. *PaperPrint* effectively creates a digital 'fingerprint' of the paper and enhances the security.

The Second Part gives brief background information about locating watermarks in hand-made paper. This is followed by the description of a simple digital procedure for recording watermarks. Details are given about overcoming problems, such as the way that, not uncommonly, watermarks in books are divided over two or more pages.

The Third Part is *Case Study—Lithuania to Russia and Sweden—Cultural—The Danzig Connection,* which illustrates the further potential of watermark research. It connects the history of the first book printed in Lithuanian (the 1547 *Catechismusa* by Martynas Mažvydas) with Danzig, with Belarus and with the enormously powerful Russian family Sheremetew.

The Fourth Part concerns the only surviving fragments of the first book printed in Estonian (a Lutheran catechism), which are rightly regarded as emblems of the national spirit. The number of pages had been uncertain but is proven in this Fourth Part. In addition the evidence from the watermark suggests a paper connection with Martin Luther.

The Fifth Part concerns the first book printed in Finnish (*ABCKiria* by Mikael Argicola) and uses the evidence from the watermark to question the accepted date, to recommend another and to illuminate the composition of book.

The Downloads section includes links to two items of supporting material. The first link shows how an octavo is made from one sheet of paper. The second link reveals the watermark found in the Vilnius copy of the above-mentioned first book printed in Lithuanian.

First Part

PaperPrint–Security

E. Forbes Miley III was a highly respected dealer of antiquarian maps.[1] However, on 8 June 2005 he was arrested at the Beinecke Library at Yale. Later investigation revealed that he had been involved in stealing at least ninety rare maps valued at some $3,000,000. A decade earlier Gilbert Bland had been found guilty of stealing maps worth some $500,000 from American universities.[2] In 2020 Michael Vinson's book related the full story of the dramatic life and death of the Texan Johnny Jenkins,[3] revealing how the man who had once achieved the largest rare book deal of the century was also a book thief. Jenkins was also suspected of forging items including a copy of *The Texas Declaration of Independence.* At the time of his death—his body with a shot to the back of the head, was discovered in a river—he was under suspicion for arson, forgery and liable for huge gambling debts. These three cases are not exceptional. The known extent of the problem of thefts is international. Thieves include priests, who were defrocked and sentenced to a year and a half in prison,[4] and insiders such as a former Bibliothèque Nationale de France conservator of Hebrew manuscripts.[5] Thefts continue up to the present day.[6] In addition, the full extent of the problem is unknown, partly because it does seem that the default attitude of those who know they have been targeted is reticence if not silence. Also, by definition, no one knows what items have been stolen if no theft has been noticed.

Clearly there is a pressing need to protect items on hand-made paper. The security of such items can be enhanced by the *PaperPrint* procedure, which was devised by the author for use at the Lindley Library of the Royal Horticultural Society. The first book there to have every page imaged using front lighting and back lighting was the 1526 *Grete herball.* Subsequently selected pages of a large number of other early printed books at the same library were imaged for research and for security enhancement. A full set of *PaperPrint* images was archived with that library. In addition, thanks to support from the Paul Mellon Centre for Studies in British Art, other copies of the *Grete herball* in London, Oxford, Texas, California and Pennsylvania have been imaged using the same procedure.

Method

The following image shows the imaging system used to capture the front-lit and the back-lit images required for a *PaperPrint* file.

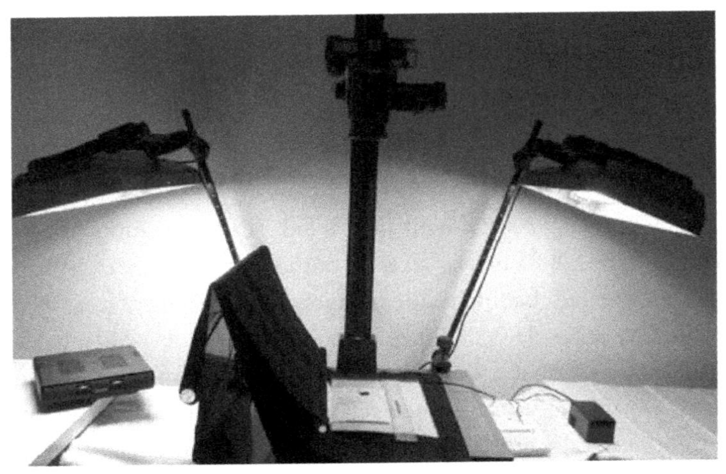

The camera is mounted on the column with the twin front lights switched on so as to illuminate the book in the cradle. The electro-luminescent light sheet is under the page to be imaged. The pink coloured light sheet (switched off here) is 1 mm thick and remains cool when switched on to give a bright uniform white light. Note the mm scale placed on the light sheet to the right of the page.

PaperPrint calls for a front-lit image:

Front-Lit Van Kouwenhoorn Title Pieter van Kouwenhoorn / RHS Lindley Collections.

and a back-lit image to be captured:

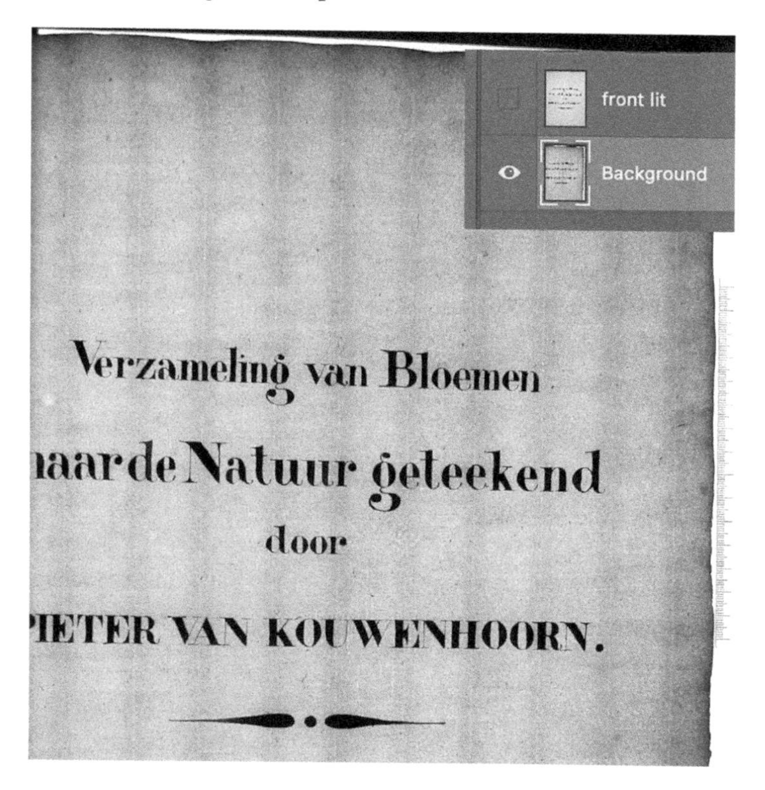

Back-Lit Van Kouwenhoorn Title Pieter van Kouwenhoorn / RHS Lindley Collections

Both images are captured under the same conditions. There should be no displacement of camera or subject so that when the images are archived as two layers on one digital file they are perfectly aligned. The images above have been annotated to show the layers on which they were archived.

The first consequence of creating that single file is that the equivalent of a digital 'fingerprint' of that paper item is available. In the event of theft and recovery the *PaperPrint* file provides indisputable evidence of provenance. Another consequence of creating the *PaperPrint* file is that a valuable research resource is created. First example—any watermark is recorded, as in the Van Kouwenhoorn Title page above. The following image shows the watermark plus a mm scale. The scale is taken from the original back-lit image and placed on its own separate (Photoshop) layer. Once on that layer the scale may be moved thus allowing accurate measurements of chain lines and laid lines for instance. The resultant (Photoshop) image has three layers—the front-lit image; the back-lit image; the mm scale.

Van Kouwenhoorn Title Pieter van Kouwenhoorn / RHS Lindley Collections

Second example—it is possible to so manipulate the two images that hidden data is revealed. Third example—it is possible to so manipulate the two images that unwanted data can be digitally 'removed' thereby revealing watermarks. The following (greatly reduced) front-lit image is of a 1695 Copper Engraving in the author's own collection.

The following image shows the centrally placed Isle of Mann by back light:

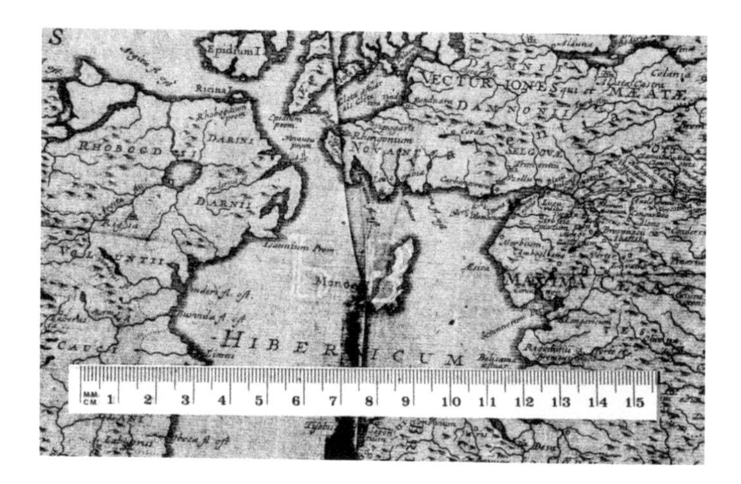

The chain lines, the laid lines and items such as specks and flecks in the paper are clearly visible. The distribution is unique to that item—a digital 'fingerprint' has been created.

Secondly, use of graphic handling software allows the front-lit and the back-lit images to be so manipulated that overprinting may be digitally removed or at least reduced. Typically this is done, with the two grey scale images being archived as one data file with the front-lit image on the upper layer and the back-lit image on the lower layer. First the shades of grey in the upper layer are inverted, so that a 'negative' image is created. When the opacity of that upper layer is reduced, typically to 50%, any data on that layer cancels out the same data which is also on the lower layer. Data which is not on both layers eg a watermark, is unaffected and free from overprinting. This is shown in the following image:

The Lindley Library and Other Examples of *PaperPrint* in Use

The Lindley Library at the Royal Horticultural Society (RHS) archive *PaperPrint* files of selected items. Here are excerpts from a presentation by Dr. Brent Elliott, formerly of the Lindley Library given at Gresham College in May 2013:

RHS Lindley Collections

The Royal Horticultural Society (RHS) is the world's largest horticultural society—membership passed the 400,000 mark in 2013—and it also has the world's largest horticultural library. In 2011 the RHS Lindley Library was given Designated status by the Museums and Libraries Association as a collection of national and international importance.

Dr. Brent Elliott described the collections and the work of the RHS Lindley Library and showed that the Library plays a key role in the preservation and study of the United Kingdom's horticultural heritage. It is the primary collection for the study of garden history in all its aspects.

See his Conclusions with regards to imaging watermarks here[7]:

> *In the course of this work, I am proud to report the Library's role in pioneering a contribution to the protection of antiquarian books. As with any collection of early printed books, the Library contains a variety of types of early paper, in many cases with watermarks. At the end of the last century, the Library sponsored a project on the imaging of watermarks, conducted by Dr Ian Christie-Miller. The analysis of the paper and watermarks in early English and French books resulted in*

> *the discovery that the older the book, the more disparate the sources of paper: since there were no paper mills in Britain in the sixteenth century, British printers tended to stockpile paper wherever they could find it, and upwards of fifteen different types of paper could be used in the production of a single book. In the course of this research Dr Christie-Miller developed his system of 'PaperPrint' identification for books printed on handmade paper. Take the title-page, and one or two other pages selected at random (so a thief does not know what to remove); photograph them using both reflected light (for easy identification) and transmitted light (to show the paper structure). Handmade paper always had imperfections and inclusions, and in no two copies will these be in exactly the same places on the page; in no two copies will the pieces of type occupy exactly the same positions in respect to the chain-lines in the paper. The result is, for an antiquarian book, the equivalent of a fingerprint: if the book ever disappears, and there is uncertainty over whether a recovered copy is the correct one, the PaperPrint will allow for an unambiguous identification. This has been a brief and inadequate account of the collections and the work of the RHS Lindley Library—but, I hope, sufficient to demonstrate that the Library plays a key role in the preservation and study of this country's horticultural heritage. It is the primary collection for the study of garden history in all its aspects.*

By way of example from the Lindley Library, Royal Horticultural Society, here are front and back-lit images of one of their treasured items, as archived as one *PaperPrint* multi-layered file:

Van Kouwenhoorn, 29 Pieter van Kouwenhoorn / RHS Lindley Collections

Here are enlargements of the insect by front lighting and by back lighting:

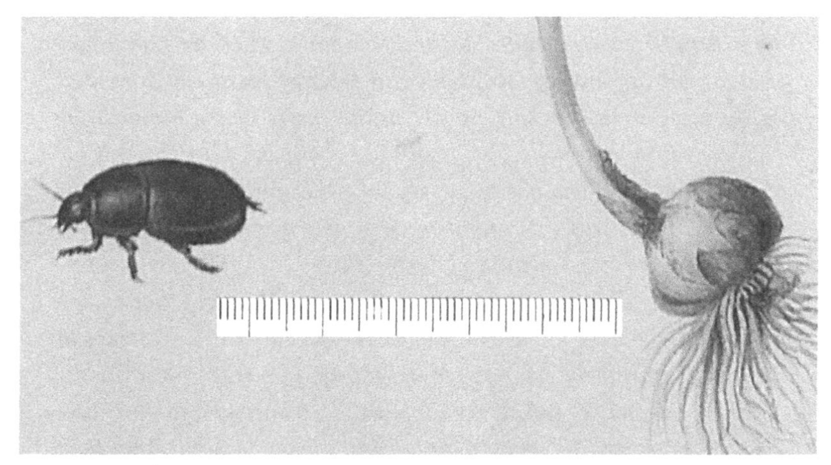

Van Kouwenhoorn, 29 Pieter van Kouwenhoorn / RHS Lindley Collections

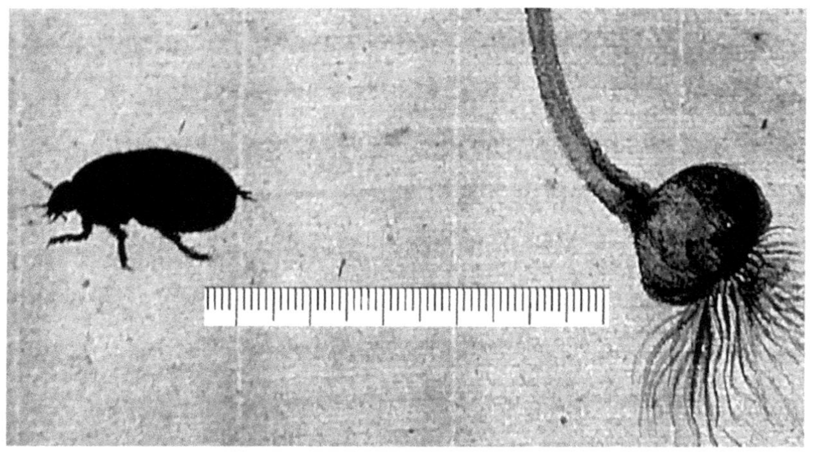

Van Kouwenhoorn, 29 Pieter van Kouwenhoorn / RHS Lindley Collections

The distribution of chain lines, laid lines and impurities in the paper is unique.

The first library outside of the United Kingdom to hold a full set of the multi-layered *PaperPrint* files was Marsh's Library in Dublin for their two copies of the rare Cyrillic *Treatise on the Sacraments,* 1657 by Kosov.[8] The following images show two of the layers of the *PaperPrint* file of B ii recto:

ѡ Та́йнахъ. ї҃

Тѣломъ и҆ Кро́вїю Хрⷭ҇то́вою быва́н. Потре́тє, и҆жъ в и҆ншихъ Сакраме́нтахъ, Мате́рїѧ не ѿмѣнѧ́етсѧ натꙋю рѣчъ ко́торꙋю зна́читъ. Напри́кладъ: Вода̃ в Крещє́нїи, зна́читъ ѿдроже́нє Дꙋ́ховное, а̑нє ѿдъмѣнѧ́етсѧ в ѿное ѿдроже́нє. в Єѵхарнстїи за́сь, ѿдменѧ́етсѧ є҆ство Хлѣба и҆ Вина, в є҆ство Тѣла и҆ Кро́ви Хвы. На ѿста́токъ и҆та́ѧ ро́зница є҆стъ, и҆жъ и҆ншихъ Сакраментѡвъ са́мъ ꙋ҆жива́ти немо́жетъ Їере́й, напри́кладъ: немо́жетъ се́бе ѿкрестити, Мѵ́ромъ пома́зати, а̑пожыва́ти самъ тогѡ Сакраме́нтꙋ Єѵхари́стїи мо́жетъ.

Вопро́съ,

Што называ́етсѧ ѿсо́бами Хлѣба;

Ѿвѣтъ.

Называ́ютсѧ примѣ́ты Хлѣбовые, ꙗ҆кѡ то Квасно́сть, Смакъ, Окрꙋ́глость, Бѣлость, и҆ про́: Такъ тежъ ѿсо́бами Вина̃, Называ́ютсѧ примѣ́ты є҆гѡ̃, ꙗ҆кѡ то Кваснѡсть, Со́лодкость, Сма́къ, и҆ про́чаѧ:

Вопро́съ.

Ѿмѣнѧютъ лисѧ тые ѿсо́бы Хлѣба и҆ Вина̃, в Тѣло и҆ Кровь Хвꙋ;

В б   Ѿвѣтъ

Courtesy of Marsh's Library

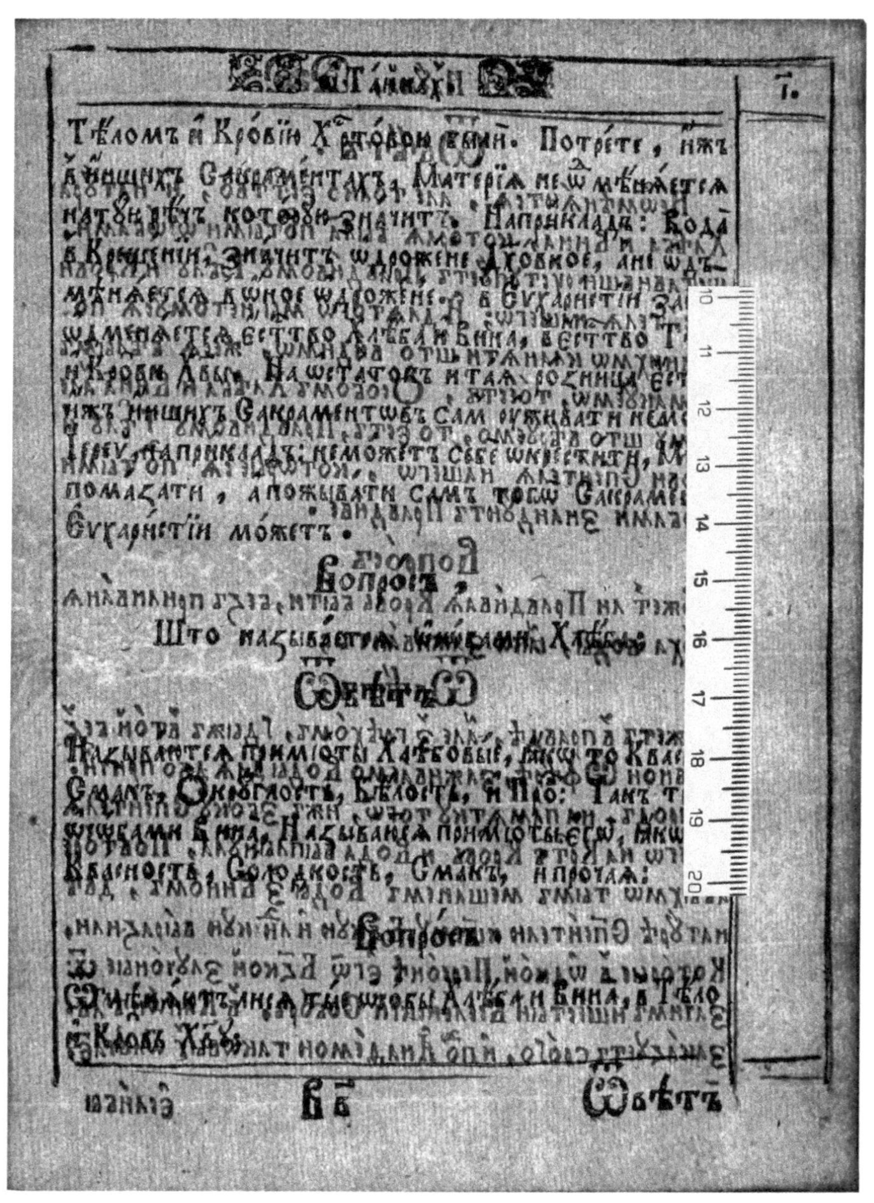

Courtesy of Marsh's Library

Part of the watermark can be seen half way up the left side of the page. It is a Foolscap. Despite the place of printing—Kiev—the watermark is remarkably similar to one found in paper used, in Amsterdam, by Rembrandt dated 1648 as shown below:

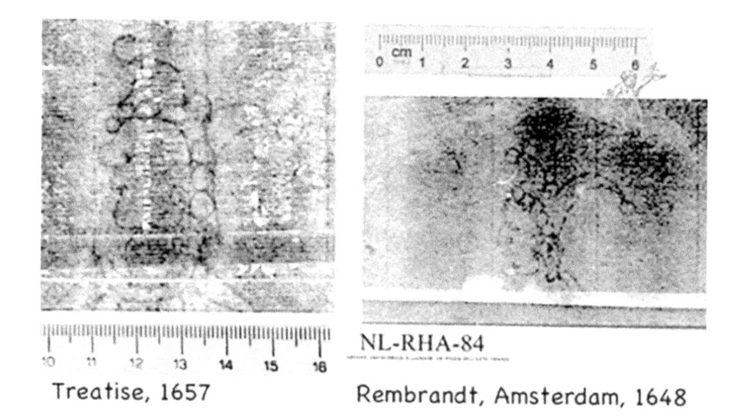

Treatise, 1657 — Rembrandt, Amsterdam, 1648

This similarity was recorded by the author for the British Library on-line Notepad[9] and shows that the abovementioned imaging processes not only enhance security but also reveal surprising historical connections.

In November 2013, at a Sotheby's, New York sale, a copy of the 1640 *Bay Psalm Book* was sold for $14,165,000, making an auction record.[10] That copy of the book (H.21.15), along with another copy (H.21.14), had belonged to the Boston Old South Church and had been in the safe keeping of Boston Public Library.

H.21.15 was sold to Mr. Rubenstein, the philanthropist, co-founder and co-executive chairman of American private equity firm The Carlyle Group. Mr. Rubenstein planned to share it with the American public by loaning it to libraries across the country, before putting it on long-term loan at one of them. H.21.14 remains at the Boston Public Library. Full sets of *PaperPrint* images of both those books have been archived with the holders.

Second Part

Paper, Pages, and Finding Watermarks

The initial production of paper in Europe relied on eastern techniques. Around 1200 the paper produced in Spain was based on Arab techniques. It was thick, made of material which was poorly beaten and without watermarks. The earliest known piece of paper in England is in Hereford Cathedral and is thought to be a Spanish/Arab paper. It was used for a letter written from Avignon around 1308.[11] Italy became an early source of paper with more northerly countries following. England was a latecomer. There is some evidence that there was a paper mill in Hertford in 1490, but it did not continue.[12]

The central activity of making paper by hand was the skilled process of immersing the mould into the vat of wet 'stuff' which was prepared, typically, from beaten rags. The mould would then be raised so that water could drain through the wires thereby beginning the drying process, as in the following picture:

The typical mould shown above is a wooden rectangle with a base of wires.

It has been noted above that the oldest known piece of paper in England has no watermark. Watermarks are due to the presence of wire(s) in the mould which inevitably means that the paper is thinner and so more light passes at those places. The French word for watermark is 'le filigrane' derived from the word for wire. It is therefore clearer than the English word 'watermark'. The major wires, known as chain lines, in the mould, generally, ran parallel to the shorter side. In the image below the chain lines are vertical. Other, lesser, wires running at right angles to the chain lines are known as laid lines. In the image below the laid lines are horizontal. In addition, distinctively shaped wires came to be sewn to the mould. Sometimes those wire shapes were emblems (unicorns for instance). Sometimes dates were given. Sometimes the name of the mill was given and so on.

The following image shows a corner of a mould, with the watermark wires, unusually, placed near the corner.

© Copyright Simon Barcham Green 2020

This mould produced paper known as 'Charles I' as shown by the watermark wires above and more clearly below:

© Copyright Simon Barcham Green 2020

This well-used mould was developed by Jack Barcham Green (note the JBG initials in the shield) in the latter part of the first half of the 1900s and was in use until 1975. Jack Green had written the letters backwards with a pen to make them look 'ancient' and the mould maker copied the drawing. The superior quality of such moulds is evident from the decades of use. The following image shows a 'Charles I' sheet by back lighting:

Photograph by Ronald M. Bodoh

A recent image of the mould shows the date '1975' as below:

The upper part of the image above shows the top right corner of the mould. The lower part shows the image cropped, rotated, flipped and processed. The sewings are evident. The damaging effects of cleaning and aging of the mould would be evident in the paper produced and can prove valuable in establishing the relative ages of papers from the mould. For a case study of the use of watermarks to assess dating see Fifth Part Composition and Dating.

The mould in use by the papermaker pictured in action above is typical in size. In contrast the 'Charles I' mould is an exception in that it is double, so that two sheets may be made at the same time, as may be seen in the following image:

© Copyright Simon Barcham Green 2020

Each page measures 465 mm by 590 mm (about 18.25 inches by 23 inches). The mould itself is exceptionally wide at just under 50 inches in width (about 1260 mm). The mould is therefore unwieldy.

Any sheet, when folded once, produces two leaves (four pages)—a bifolium. After both sides have been printed, but before being folded, the left-hand side of the sheet has page 4 on the left (backed by page 3). The right-hand side of the sheet has page 1 (backed by page 2). Notwithstanding the Charles I example given above, customarily the watermark emblem was on the middle of the right-hand half of the mould. It is a feature of moulds and of papers that every aspect has exceptions, as above.[13] For a folio the watermark is invariably found in the centre of one of the two pages making it easily seen. Sometimes the other half of a sheet would bear another mark—the countermark. Predictably there are variant positions, as above.

Consideration will now be given to books smaller than folios.

The next smaller, known as quartos, have four leaves (eight pages). This is evident from the following illustration, which shows the numbers of four pages on one side of a printed sheet:

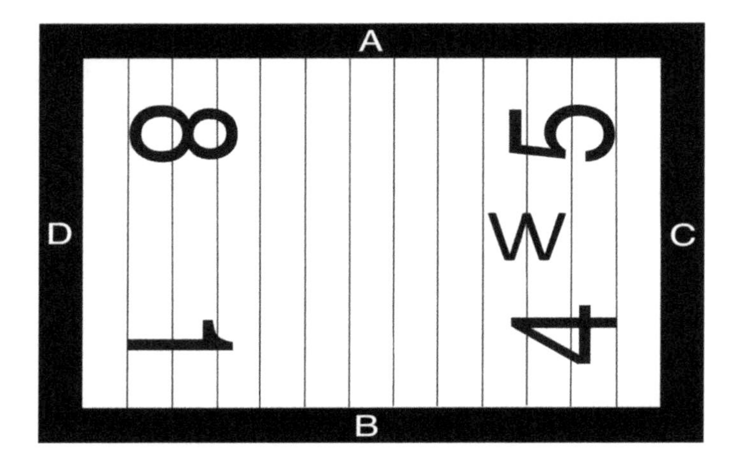

The vertical lines show the typical distribution of chain lines. The 'W' shows the usual position of the watermark. In the case of a quarto the large numbers in the illustration above show the page numbers for one side of a sheet of paper.

Page 1 is backed by page 2.

Page 4 is backed by page 3.

Page 5 is backed by page 6.

Page 8 is backed by page 7.

The following steps are needed to assemble the gathering of (almost all) quartos:

1 Fold the left-hand side of the sheet along line AB so that pages 1 and 8 are at the front.

2 Fold the folded sheet again, this time along line CD so that page 1 is still at the front.

Typically in a folio the watermark is central to the leaf, but in a quarto it would be divided over two leaves. It lies between page 4 (backing page 3) and page 5 (backing page 6).

Of course the printer might have started with the paper rotated 180 degrees. In this case the watermark would be divided between page 1 (backing page 2) and page 8 (backing page 7). It will be seen that the chain lines in this completed quarto run horizontally This is a general, but not invariable, feature of quartos. Typically, due to the folding, the watermark is not only divided between two pages but also 'disappears' into the gutter. This presents considerable difficulty for research. A similar division of the watermark occurs in the next smaller book size—the octavo. That smaller book size will now be discussed.

Typically for a quarto the chain lines, as shown above, run horizontally. Typically for an octavo they run vertically as they do for a folio. This is evident from the following illustration:

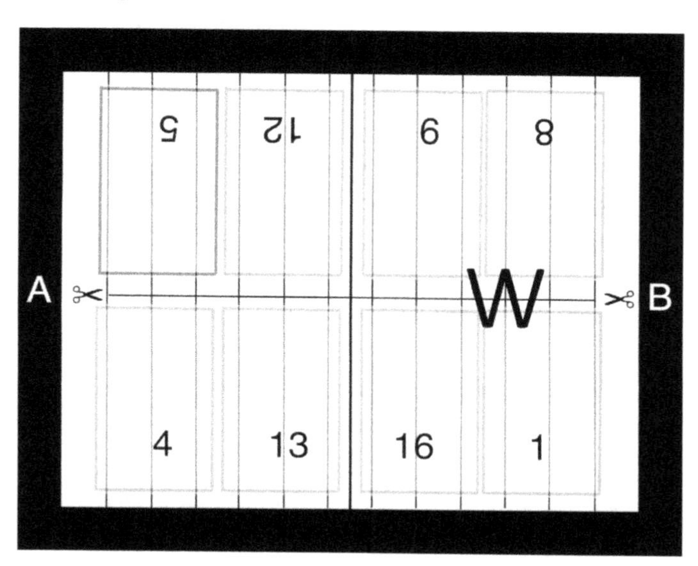

Not uncommonly the watermark is divided over four leaves. In this case page 1 (backing page 2), page 16 (backing page 15), page 8 (backing page 7), and page 9 (backing page 10). This is an undoubted difficulty when researching watermarks, but it is by no means insurmountable as shown below in the *Third Part—Case Study—Lithuania to Russia and Sweden—Cultural—The Danzig Connection.*

The octavo is made by following these steps:

1. Cut horizontally as shown along AB. This creates two strips.
2. Take this lower strip and fold it so that 1 and 16 show.
3. Fold this lower strip once more so that 1 shows.
4. Take the upper strip and rotate it 180 degrees.
5. Fold this upper strip so that 12 and 5 show.
6. Fold this upper strip once more so that 5 shows.
7. Insert this folded upper strip into the middle of the folded lower strip.

A movie showing this process is available (see Octavo Assemble in the DOWNLOADS below). It is possible to recreate an octavo by printing out the two sides of a single sheet with the eight pages correctly disposed on each side.

To do this see Octavo Both Sides in DOWNLOADS below. This recreates the sheet as originally printed. The images in the download show Gathering A from the Vilnius copy of the first book printed in Lithuanian. The book is Martynas Mažvydas, CATECHISMVSA prasty Szadei, Weinreich, Jan (ca. 1490-1560) (Kaliningrad, 1547).

The difficulties of finding and researching watermarks in even smaller books are, naturally, greater. For fuller information about printed sheets in duodecimo, including possible placing of watermarks, works such as *An Introduction to Bibliography for Literary Students* should be consulted.[14]

Most early printed books have watermarks. Some have the same watermark throughout. Others have multiple watermarks. For instance, the Royal Horticultural Society copy of the 1526 *The Grete herball* has a great variety of French watermarks such as unicorns (39), hands (6), shields (14) and Gothic letters P (21). Some watermarks in that book are single, but others appear more than once. By way of example, the image below shows 6 of the unicorns.

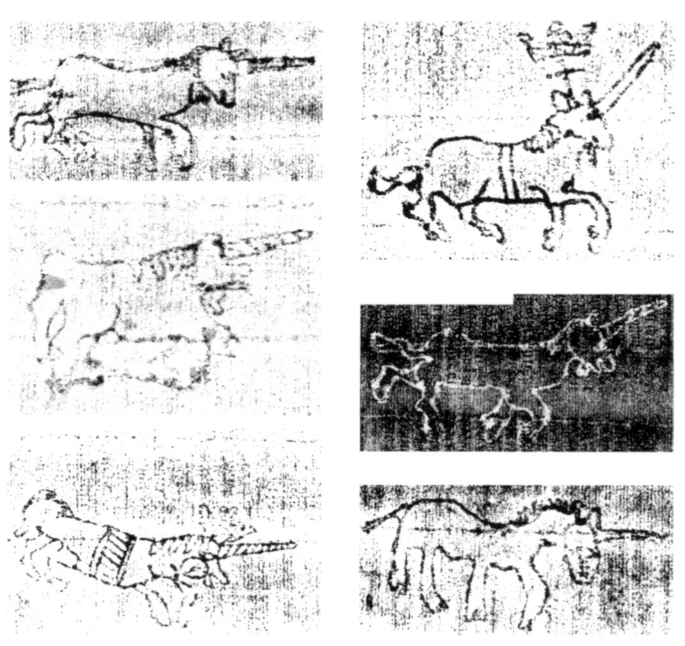

RHS Lindley Collections

The Grete herball is a folio bound in sixes. This means that 3 sheets were used for each gathering and so there would, normally, be 3 watermarks—one at leaf 1 or leaf 6; one at leaf 2 or leaf 5; one at leaf 3 or leaf 4. In each gathering there would be 6 leaves giving 12 pages but only 3 watermarks. There is an exception at gathering L which has 4 watermarks. There are Gothic P watermarks at leaves 1, 5 and 6 plus a unicorn watermark at leaf 4. Clearly the copy of *The Grete herball* held at the Lindley Library of the Royal Horticultural Society is a composite and not the original single copy.[15] This small example shows how watermarks can provide unexpected discoveries about the history of a particular book.

Imaging Procedure

The earliest images of watermarks were sketches, as found for instance in Briquet's monumental *Les Filigranes.*[16] More accurate records could be made by tracing. Various radioactive sources have also been used but these have significant health and safety disadvantages. The resultant images do have the significant advantage of not recording text so that watermarks are not obscured. Infrared has the similar ability to 'ignore' some types of ink used for text. Conventional cameras are not sensitive to infrared. However most may be specially adapted by the removal of the internal infrared cut-out filter.

In order to find and to illuminate the watermark the simplest method is to use an electroluminescent light sheet. These 1 mm thick, cool, mains powered sheets provide steady even white light. The sheets are available in all sizes. The light sheet can be easily placed under the page to reveal the watermark and to allow the image to be captured.

The Grete herball 1526, as noted above, is a folio bound in sixes. The watermarks appear in the middle of the pages, so finding the watermark is easy, although the text, albeit in 2 columns, is more than likely to obscure it, at least in part.

Watermarks in smaller books are difficult to locate partly because they are in the gutter of the binding. Light sheets are invaluable because they can be easily inserted into the gutter of the book. The recommended way of digitally 're-assembling' such divided watermarks is described in the next section.

Image Processing and Archiving

Firstly it is recommended that handwritten notes (pencil) be made for each imaging session. Suggested items to be recorded are:

Date of imaging session
Place of imaging
Summary of imaging system used

Details of the books to be imaged e.g., title, accession number, size
Running record of the camera exposure number—Column 1
Running record of the page /leaf/gathering number—Column 2
Running Record whether recto or verso—Column 3 or 4
Running Record whether front-lit or back—Column 5 or 6
Generous space beside each Running Record for later comments and notes eg sketches of watermarks—Column 7 or 8

A typical layout for such notes is shown below:

Date 16/12/2018
Place Riga, NL
Imaging system used Portable Light Sheet / Reflector/Canon 5D
Book / paper Title Luther – Der kleine katchismus
Accession L 68-2/72

Image No.	PAGE	R°	V°	FRONT	BACK	spare	COMMENTS
IMG_0012	D	✓		✓			
IMG_0013	D	✓			✓		Part Countermark
IMG_0014	D2	✓		✓			
IMG_0015	D2	✓			✓		Part Laurel watermark

To print a blank version of the Notes see Notes in DOWNLOADS below.

At the end of each imaging session the notes themselves should be imaged and those images filed along with the others. One reason is for forensic purposes. Another is that in the event of an error, such as a page being omitted, it is far easier to remedy the error. Second, it is recommended that a mm scale be imaged alongside the book. The image of the scale should be archived, as described below, together with the file of the front-lit and the file of the back-lit image. Third, it is recommended that the files be recorded in both RAW and another format such as JPEG. The advantage of RAW is that fuller data is recorded than for the other format. This advantage has forensic potential and is carried forward when changes are made to the original RAW file. The disadvantage of RAW is that the file size is considerably greater than the other format.

Image processing software is used to 'reassemble' watermarks as follows. For a quarto, two images are selected. With reference to the quarto image given above, back-lit images of pages 4 and 5 are chosen and placed one beside the other so as to reveal as much of the watermark which is in the gutter. (The same result would be achieved if pages 3 and 6 were chosen.) If the printer had rotated the paper 180 degrees, then pages 1 and 8 or 2 and 7 would be concerned. It is a feature of quartos that the sum of the page numbers involved is 9. For an octavo, four images are selected. With reference to the octavo image given above, back-lit images of pages 1, 8, 9 and 16 are chosen. Pages 8 and 9 (or 1 and 16) are rotated. All four are moved so as to reveal as much as possible of the watermark in the gutter. (The same result would be achieved if pages 2, 7, 10 and 15 were chosen.) If the printer had rotated the paper 180 degrees then pages 4, 5, 12 and 13 or 3, 6, 11 and 14 would be concerned. It is a feature of octavos that summing the page numbers involved gives 34. Summing the first pair gives 9. Summing the second pair gives 25.

Image processing software such as Photoshop or the (free) GNU Image Manipulation Program[17] is used for archiving and processing the images. The Layers facility is needed for archiving separately the front-lit image, the back-lit image and the image of the rule all together but on their own layers. Below are some basic ways of processing those images.

To make measurements: the rule may be copied from its layer and pasted onto another layer, such as the layer holding the back-lit image. The rule can be rotated and moved at will so as to make measurements such as the spacing of chain lines.

The procedure to remove unwanted data to reveal a watermark has been given above. The same procedure can be used to reveal information which would otherwise be hidden. The front-lit image of a page may be digitally 'subtracted' from the back-lit image of the same page, thereby revealing data on the under surface. One simple way of achieving this 'subtraction', as referred to above, is by selecting Mode Greyscale, selecting and 'inverting' the upper of the two image layers and reducing this upper layer opacity to 50%. Any data common to both images disappears. Any data not common to both, such as a watermark or any data on the undersurface, is evident, albeit reduced in intensity. This revealed data may then be enhanced. This procedure has also been used to reveal information on the undersurface of an item with a translucent backing. Here are one example from the Victoria and Albert Museum (V&A) and one from Sir John Soane's Museum collections.

The first example is from the V&A. 48.B.12 manuscript pages containing changes proposed by Charles Dickens to *The Chimes.* The autograph pages had been carefully mounted on translucent backing and bound rendering it impossible to read the author's proposed corrections on the undersurfaces. The abovementioned process allowed hidden text to be revealed as follows.

Here is an image of leaf 54 as seen by front lighting (on the left) and by back lighting (on the right):

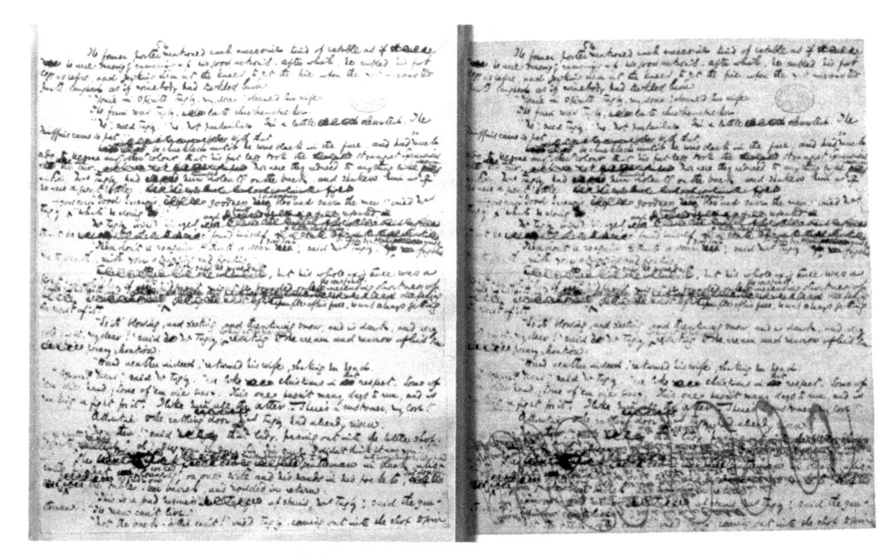

© Courtesy of the Victoria and Albert Museum, London

The imaging process calls for two images of each sheet to be captured and processed. One image is with conventional front lighting. This is shown above on the left. The second image is captured, without moving the page, with back lighting. This is shown above on the right. The backlighting is from a 1 mm thick electroluminescent sheet placed under the leaf. The front-lit image records front surface writing (which is not needed). The back-lit image records the same front surface writing, as well as writing on the back surface (which is needed). The task is to subtract the front-lit image from the back-lit image, thereby leaving only the writing on the back surface.

The process relies on making writing which occurs on both images cancel itself out. The first step is to prepare one data file (typically with Photoshop) which holds both images. The two images are placed on their own layers. The resultant single (Photoshop .psd) file thus records both images perfectly superimposed. Use Photoshop as follows:

1. Under Image / Mode choose Greyscale.
2. In the 'Layers' control panel select the upper layer.
3. Use Image / Adjustments / Invert. This makes dark greys light and vice versa.
4. In the 'Layers' control panel reduce the opacity of the upper layer until all unwanted writing disappears = self-cancelling.

5. Flatten the layers. This means that there is now only one image / layer. Unwanted data has been removed by self-cancellation.
6. Adjust levels so as to improve appearance.

Here is the result for the lower part of leaf 54:

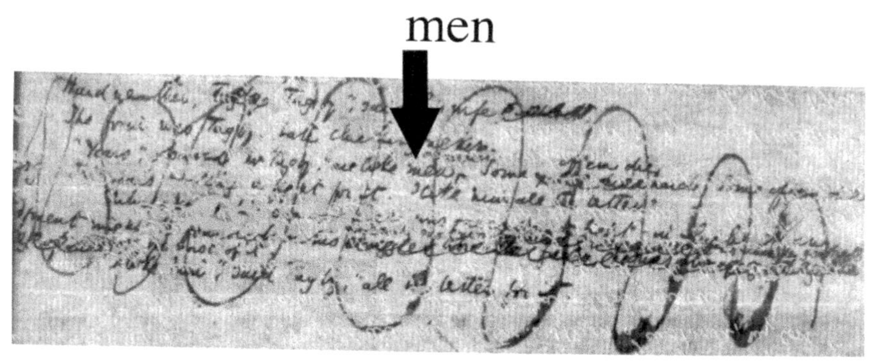

© Courtesy of the Victoria and Albert Museum, London

Note that the image above starts with 'Hard weather ..' and ends with 'all the better for it.' The word 'men' is indicated and is discussed more fully below. The printed text is found in the Fourth Quarter of *The Chimes*:

> 'Hard weather indeed', returned his wife, shaking her head.
> 'Aye, aye! Years', said Mr. Tugby, 'are like christians in that respect. Some of 'em die hard; some of 'em die easy. This one hasn't many days to run, and is making a fight for it. I like him all the better. There's a customer, my love!'

The undersurface of the manuscript (as shown above) has numerous differences from the printed text. In the third line, for instance, there is the word 'men' (as shown above) as opposed to 'Christians' in the printed text. Here, by way of comparison, is the front-lit portion, which is also in the printed text:

© Courtesy of the Victoria and Albert Museum, London

The full set of images is now held at the V&A and is available to researchers. The results of the imaging procedure demonstrate how such revelations can lead to further questions. For instance why did Charles Dickens, with his well-known social sensitivities, choose to change the wording? Evidently he wavered.

The second example is from Sir John Soane's Museum in London[18] and concerns that museum's holding of architectural drawings by Robert Adam (1728-92), Scotland's greatest architect.[19] Among the museum's treasures are many albums containing drawings drawn and stuck down by Robert Adam. Thus in many cases only the upper surface can be seen and studied. One example is Drawing 40 in volume 55 of the Adam collection. The drawing is secured at the four corners and is shown here.

Courtesy of Sir John Soane's Museum

Because the drawing was secured only at the four corners, it was possible to insert the light sheet between the drawing and the opaque backing. In this way it was possible to illuminate the drawing from the back thanks to the use of the 1 mm-thick electroluminescent light sheet inserted at different angles. Three images were taken, the light sheet being manoeuvred so as to avoid the four contact points. The three images were then combined. Use of the abovementioned digital subtraction process was then carried out. The result is shown here.

Courtesy of Sir John Soane's Museum

Under Neptune's outstretched hand there is a clear watermark 'AG', which has been enhanced and is shown next.

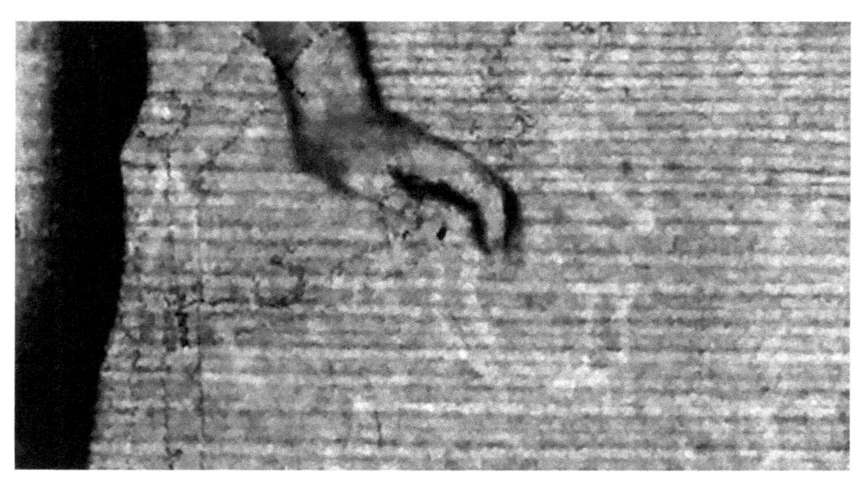

Courtesy of Sir John Soane's Museum

These examples from the V&A and Sir John Soane's Museum conclusively show the ability of back lighting to reveal data which would otherwise be hidden.

Archiving of the images is most conveniently done with a small datafile in a database. By way of example, with Filemaker Pro a folder is created such as 'Project XXX'. That folder would contain, firstly, the small datafile 'Project XXX.fmp12'. This small datafile would be set up with whatever fields are desired, such as Separate Text fields created for information e.g. page number. Separate Container fields are created for images. In order to keep the datafile small, it is essential that files of images not be entered into any Container field. The solution is that the 'Store only a reference to the file' option be selected. Thus only a 'signpost' is entered into the Container field. All images should be archived into one folder, such as 'Project XXX Images'. That folder, 'Project XXX Images', and the datafile, 'Project XXX.fmp12', are both stored in the folder 'Project XXX'.

The processes described above have relied on white light for front lighting and for back lighting. Increasing use is being made of other types of light, notably infrared. The use of infrared has evident advantages. The most obvious is that the shorter wavelength light passes through certain inks. This means that infrared effectively ignores those certain inks. The following images from Psalm 118 of the King's College London copy of the 1513 *The Psalms, followed by Sacred Canticles and the Song of Songs*[20] show how infra red can be used to eliminate ink spilled on a printed book:

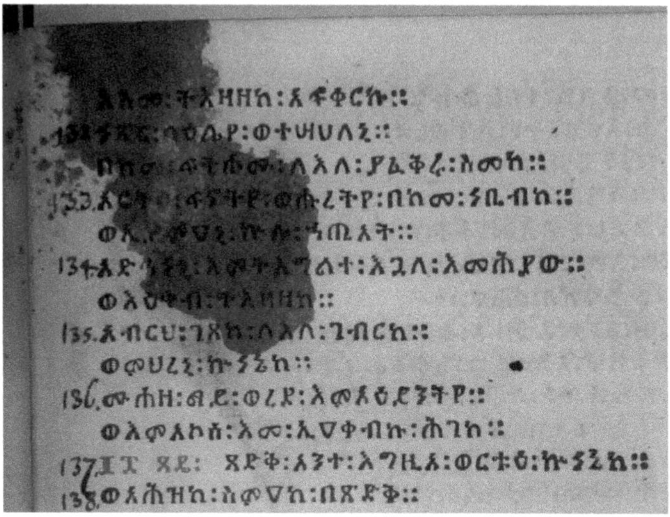

Image above taken by WHITE light King's College London, Foyle Special Collections Library

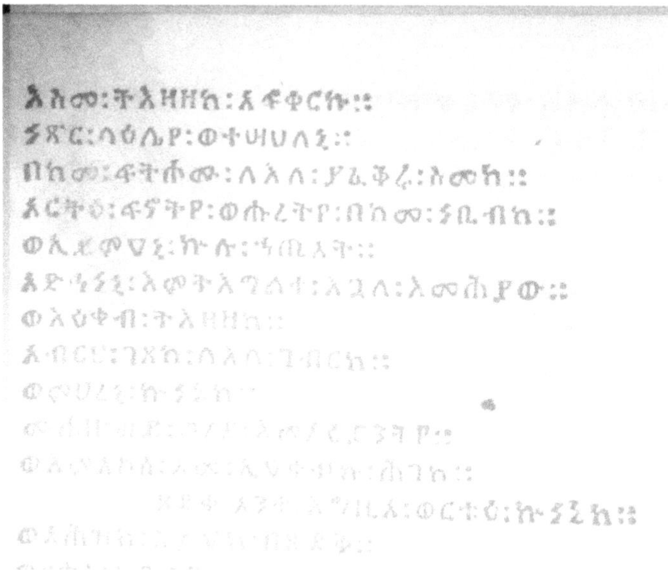

Image above taken by INFRARED light King's College London, Foyle Special Collections Library

Specialised equipment is needed for successful imaging with infrared. For the images shown above, an adapted camera and a dedicated light source were used. The Canon camera had been made sensitive to infrared by the removal of the internal infrared cut-out filter. The dedicated light source was an array of 120 light emitting diodes working in the region of 940 nanometres. The human eye and brain are not aware at those wavelengths so it is necessary not to look at the light source.

There are fuller details of suitable equipment available for imaging with white light and with infrared for both front lighting and back lighting at:

http://www.earlybook.info

Third Part

Case Study—Lithuania to Russia and Sweden—Cultural—the Danzig Connection

The author, thanks to the Vilnius University library, and with funding from the Bibliographical Society and from the Lutheran Church Missouri Synod, has been able to make extensive study of one of the two extant copies of the first book printed in Lithuanian—the 1547 *Catechismusa Prasty Szadei,* by Martynas Mažvydas. Although it was originally stated that there are no watermarks in the book, use of the Early Book Imaging System showed that watermarks are present. Subsequent research not only allowed the watermark to be identified but also led to significant and extensive cultural and historical discoveries as described below, ranging across the Baltic from Russia to Sweden.

The 1547 *Catechismusa Prasty Szadei,* by Martynas Mažvydas,[21] was printed by Jan Weinreich (ca. 1490-1560) in Königsberg, now Kaliningrad. Two extant copies of the book exist, one in Toruń[22] and one in Vilnius.[23] The name of the author is not overt: it was discovered in 1938 by Jan Sefarewicz,[24] in hidden form, as can be seen from this image of Aiiij:

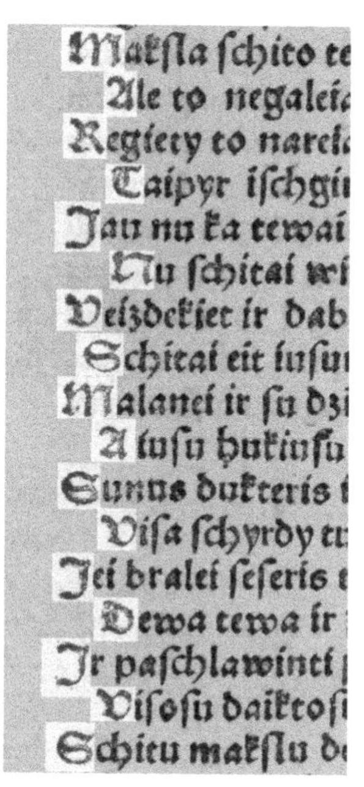
Makſla ſchito te
Ale to negalei
Regiety to narei
Taipyr iſchgi
Jau nu ka tewai
Nu ſchitai wi
Weizdekiet ir dab
Schitai eit iuſu
Malanei ir ſu dzi
A tuſu hukiuſu
Sunus dukteris
Wiſa ſchyrdy te
Jei bralei ſeſeris
Dewa tewa ir
Ir paſchlawinti
Wiſoſu daiktoſi
Schitu makſlu de

Courtesy of Vilnius University Library

This Lutheran Catechism is notable for having the music notation for many pages printed. All the music is available online from the endnotes in this book.[25] Although the book is mostly in Lithuanian, some parts are in Latin, such as the page 2 *Ad Magnum Ducatum Litvaniæ* and page 3 *Pastoribus et Ministris Ecclesiarum in Lithuania gratiam et pacem,* by Fridericus Staphylus, Professor of Theology and Rector of the University of Königsberg. It will now be shown how the evidence from watermarks bears further witness to the striving for a new independent order, as the Baltic countries left the Latin-dominated world and adopted the teachings of the Lutheran Reformation.

Catechismusa Prasty Szadei is an octavo. The watermark in the Vilnius copy is a distinctive crown: it was revealed, as shown below, by use of the author's Early Book Imaging System (EBIS).[26] The images below are from Gathering A. The back-lit image may be downloaded from:

https://qrgo.page.link/46D3z

The images of the eight pages on each side of the sheet-as-originally-printed were appropriately arranged, using Photoshop. The result for gathering A outer forme is as shown here:

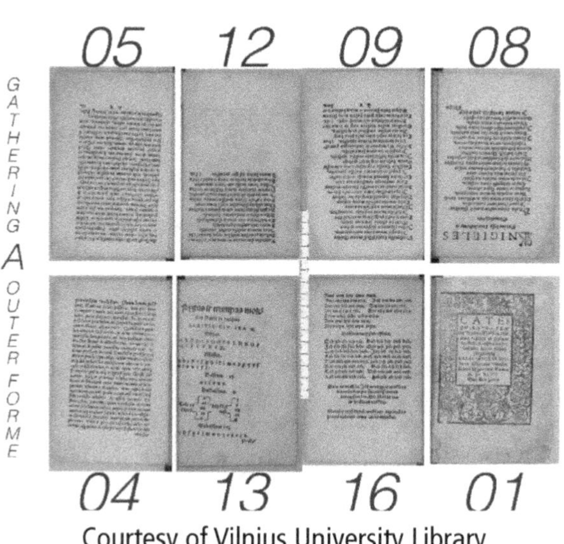

Courtesy of Vilnius University Library

The same process was applied to the back-lit images as shown below:

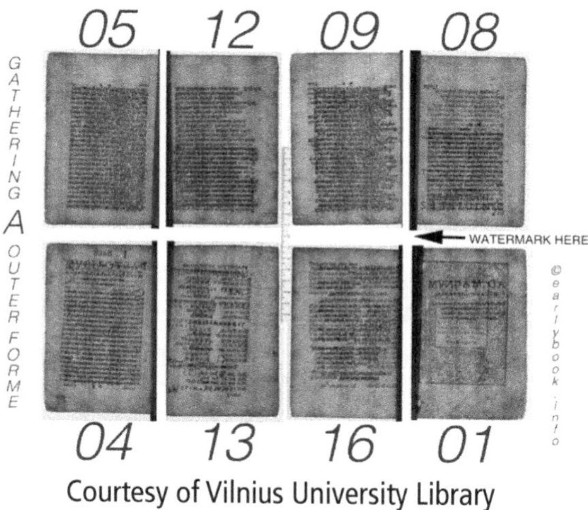

Courtesy of Vilnius University Library

As is invariably the case for an octavo, the watermark falls in the gutter of the book. In this case it is also divided over pages 1, 8, 9 and 16. As shown below, it is possible to see enough of the parts of the watermark for it to be identified as a crown:

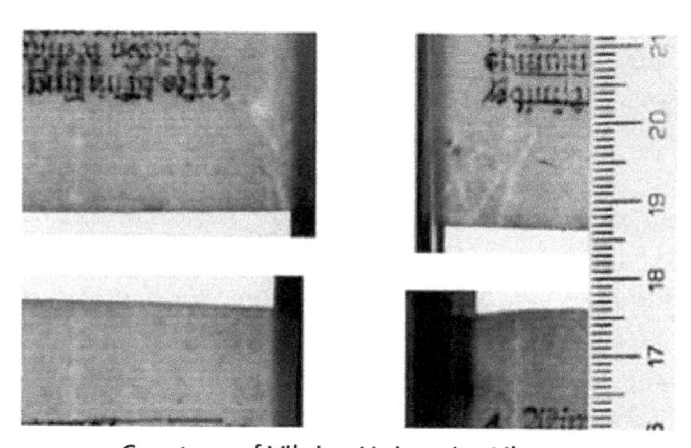
Courtesy of Vilnius University Library

Crown watermarks from the sixteenth century are very common. This crown watermark is unusual, and this is significant. The online Bernstein Memory of Paper database, which records over 260,000 watermarks, has 98 hits for 'crown' on its own in 1547,[27] and of these the most common is the closed crown, of which there are 46. There are only three of the open crown on its own, as shown above, for 1547. There is one from Wittenberg and two from

Königsberg (see below). The above-mentioned crown is not yet included on the Memory of Paper database.

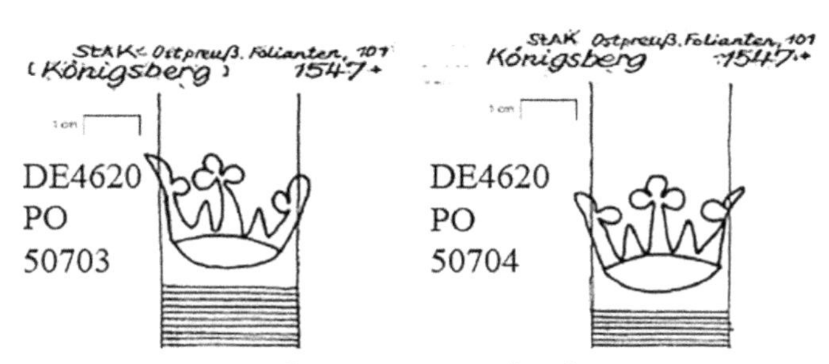

www.memoryofpaper.eu Piccard Online, Germany

A search of the entire Memory of Paper database for the same description, irrespective of year, only gives one more match. It is for Vilnius 1552.[28] The significance of the very form of the distinctive open crown is evident from seals from the Königsberg area, as shown below from Königsberg / Altstadt 1360: Löbenicht 1413: Kneiphof 1383:

The political situation is relevant. In 1525, Albrecht von Preussen (1490 - 1568) had established the first Lutheran state in Europe. Albrecht was determined for all to take a Lutheran direction, and founded the *Collegium Albertinum* in Königsberg in 1544. The university was on the island of Kneiphof and was financed by the cities Kneiphof, Altstadt and Löbenicht. (See seals above). Ford, commenting on Martyn Mažvydas, wrote:[29]

> Martynas Mažvydas (ca. 1520-1563) registered at Königsberg University under the name Martinus Moswidius in 1546. Significantly he was funded by Prince Albrecht of Brandenburg (1490–1568). Albrecht was the last Grand Master of the Knights of the Teutonic Order who, after converting to Lutheranism, created the Duchy of Prussia in 1525. This Duchy was first Protestant state. The capital was Königsberg.

Frost, commenting on the wider historical significance, wrote:[30]

> Sigismund (King of Poland, d. 1548) may have acted as midwife for the first Lutheran state in Europe (Duchy of Prussia under Albrecht von Preussen, d. 1568) but he responded quickly and decisively to stem the spread of Lutheranism in Danzig . . .

The mention of Danzig is significant. Here is an image of the Coats of Arms of Gdańsk (Danzig), Elbląg and Królewiec (Königsberg / Kaliningrad):

The similarity of the open crown to the watermark is evident. Examples of the widespread distribution, across the Baltic, of the distinctive crown and the two crosses of the Danzig arms will now be detailed.

One example of the same open style crown in the Danzig coat of arms was found by the author on a visit to Vilnius, which is 500 km to the East of Danzig. One item in the St Casimir Market was a large cooking vessel with a metal plaque at the front as shown in the following image:

The second example is a coat of arms on the wall of the main staircase of the Museum of the History of Rīga and Navigation, which some 750 km to the North East of Danzig. It is described as the 'Coat of Arms of Graf Scheremetew' (sic):

Courtesy The Museum of the History of Rīga and Navigation

The similarity to the Danzig coat of arms is undoubted. Noteworthy additions, which will be considered below, are the two lions (bearing a sceptre and an orb), the tree and the repeated crown. The German spelling of *Scheremetew* with a 'c', is to be noted. The lions have nothing in their mouths.

The third example is from Russia—Moscow is nearly 1,500 km from Danzig.

The Sheremetew family was one of the most prominent of Russian families. Their renown is reflected in the name of Sheremetyevo International Airport near Moscow.

The following image records the Sheremetew coat of arms:

From gerbovnik.ru

The basic form of the Danzig coat of arms—a crown (a different one) and the two white crosses—has been retained. Like in the Vilnius and Rīga examples there are two lions. Like in the Rīga example, the lions bear a sceptre and an orb, and a tree surmounts the crest. Intriguingly the Rīga form is an intermediary between the Danzig and the Sheremetew coat of arms as shown above. In the Sheremetew coat of arms there are two crowns neither of which resemble the Danzig, Vilnius and Rīga crowns.

I am indebted to Alexander Khmelevsky, co-moderator of the site gerbovnik.ru,

for fuller information about the Sheremetew crown. He states:

> Sheremetev: one of the most prominent clans of the Kingdom (Zarstvo) of the Russia and Russian Empire, which belonged to the General-Field Marshal Peter Sheremetev, first in Russia granted (in 1706) the title of count. Like the Royal Romanov dynasty, they derive their origin from Andrey Kobyla ('the Mare') and his son Fyodor Koshka. The great-grandson of 'the Mare' was Andrey Konstantinovich Bezzubtsev, nicknamed 'Sheremet'. The issue of Andrey 'Sheremet' was named Sheremetev.
>
> All Russian heraldry arose in the seventeenth century and was developed in the following three centuries. Therefore, in the sixteenth century, during the reign of Ivan the Terrible there could not be a question of the coat of arms of Sheremetev. The Sheremetev coat of arms is known since the end of the seventeenth century. Its prototype was the emblem of the Prussian city of Danzig (Gdańsk), in order to confirm a family legend about the origin from Prussia. The coat of arms has similarities with the arms of Danzig, Konigsberg, Elbing, and other Prussian cities, to which belonged, according to legends, the ancestors of Andrey Kobyla, Gland Kanbila and his brother

> Russinger. The modern coat of arms of the Polish city of Gdańsk (Prussian Danzig) corresponds exactly to the image on your 'converted farm vehicle'. [See the Lithuanian plaque from the St Casimir market in the text above].

A notable member of this family was Yelena Sheremetewa (d. 1587), daughter of the boyar Ivan Vasilyevich Sheremetew. She was the third wife of the son of Ivan the Terrible, who, having assaulted Yelena, killed his own son, her husband. There are various explanations for the killing: generally it is held that Ivan was incensed by the obscene clothing of his daughter-in-law, but one alternative is that it was simply a case of sexual harassment.

The Sheremetew coat of arms is very publicly evident at the magnificently restored St Petersburg Sheremetew Palace. An example of the military and heraldic prominence of the Sheremetew family is the book devoted to their armaments, by Eduard von Lenz and Count Sergy Dmitrievich Sheremetew: *Die Waffensammlung des Grafen S. D. Scheremetew in St. Petersburg . . . Mit 26 Lichtdrucktafeln*, (Leipzig, 1897). As shown in the following image, this fine book shows the Sheremetew coat of arms not only on the front cover (left, with closed crown and two stars adjacent) but also in the picture (right with closed crown and no stars) of the engraving on the barrel of a rifle made by Gibson of London.

Images of Warburg Institute copy taken by the author CGL 1200 Bay 236

As to the form of the crown itself, Alexander Khmelevsky, in a personal message dated 3 January 2019, helpfully wrote:

> This crown has no exact name—neither in Russian nor in English. Usually in Russia in the coats of arms of Sheremetev and other descendants of Andrei Kobyla, as well as in the coat of arms of Danzig (Gdańsk), it is traditionally depicted as an ancient Royal crown.
> In the blazon of Sheremetev it literally says: 'a Golden crown, i.e. the crest of the ancient Rulers of the Prussian' (in Russian «золотая корона, т.е. герб древних владетелей Прусских».)

Catechismusa Prasty Szadei, with its distinct open, Danzig style, crown, appeared in 1547. Alexander Khmelevsky also states, as noted above, that 'All Russian heraldry arose in the seventeenth and developed in XVIII–XIX centuries'. When the family Sheremetew adopted the Danzig arms the Danzig open crown was replaced by the enclosed crown shown above. The Danzig coat of arms (essentially the golden crown—some with two stars adjacent, some without stars—and the twin white crosses) was however adopted not only by the Sheremetews but also by the following families: Boborykin, Kolychev, Konovnitsyn, Lodygin, Neplyuev and Yakovlev, as shown below:[31]

Boborykin Kolychev

gerbovnik.ru

Only two retain the Danzig open crown—Kolychev and Kohovnitsyn.

Whereas the crown watermark in *Catechismusa Prasty Szadei* is the classic Danzig open crown, the Sheremetew crown (notwithstanding the Rīga example) is the closed Russian royal type.

The Sheremetew family was not only able to trace their ancestry back to the thirteenth century but maintained a position of power over many generations. The family leased out many of their estates. On 25 June 1780 Count Sheremetew gave permission to one of his 140,000 serfs—Ivan Grigor'evich

Toropov—to build a paper-mill on the River Yukhot' near the village of Kadoshnikovo. The mill was in the Rîbinsk District (formerly the Uglich District) of Yaroslavl Province. It is about 300 km NNE of Moscow. The Yukhotskoe mill shown is one of the most northerly:[32]

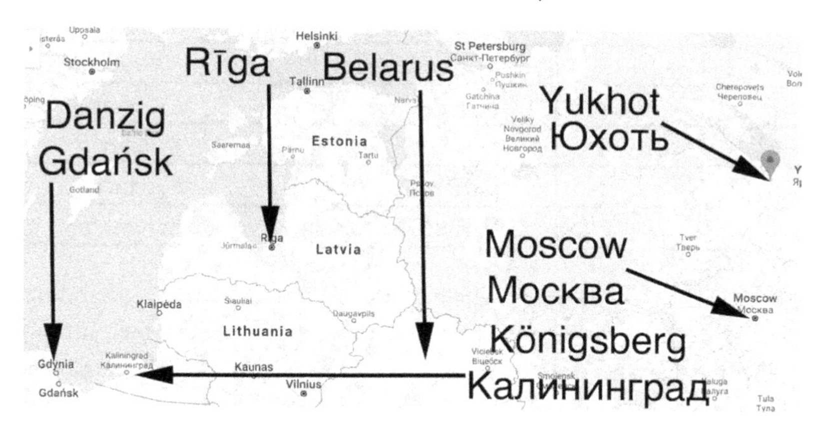

The following image shows the typical Yukhotskoe watermark:[33]

Special Collections, University of Amsterdam, hs. XXI C 19

The familiar elements of the Danzig coat of arms are evident in this unusually large watermark. The letters around the watermark are:[34]

Cyrillic Letter	Russian	English
Е	ЕГО	His
С	СИЯТЕЛЬСТВО	Grace
Г	ГРАФ	Count
П	ПЕТР	Peter
Б	БОРИСОВИЧ	Borisovich
Ш	ШЕРЕМЕТЕВЫ	Sheremetew

In the centre of the other half-sheet are the letters 'YuFST'—standing for Fabrika Soderzhatelya Toropova (i.e. Yukhotskoe mill of the owner Toropov).

The fourth example of the Danzig coat of arms watermark is from Belarus, which is some 700 km East of Danzig. These watermarks are shown below:

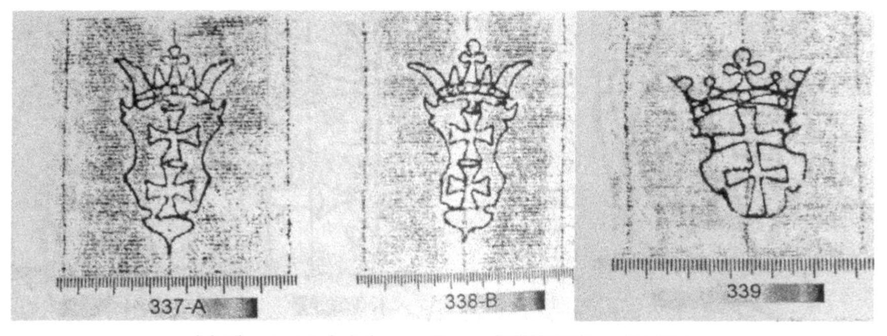

(c) The British Library Board YF.2013.a.15654.

Valeryĭ Si͡ami͡onavich Pazdni͡akoŭ, *Filihrani arkhiŭnykh dakumentaŭ Belarusi XVI – pachatku XX st*

Watermarks of archival documents of Belarus of the 16th–early 20th century/Watermarks of archival documents of Belarus of the 16th–early 20th century. (Minsk: BelNDIDAS, 2013).

The same Danzig coat of arms was also found not only to the East of Danzig but to the West. The Stockholm Vasamuseet holds a cannon from another warship, the Stora Nyckeln, which sank near Stockholm in September 1628. The cannon was cast in Poland—the date 1535 is evident on the barrel as shown below:

Courtesy of Vasamuseet, Stockholm

This date is 12 years before the publication, in Königsberg as noted above, of the first book printed in Lithuanian. The coat of arms of Danzig with the familiar open crown is shown below:

Courtesy of Vasamuseet, Stockholm

It has been shown above that the crown discovered in the Vilnius copy of the 1547 *Catechismusa Prasty Szadei,* by Martynas Mažvydas was alone—it did not have the two crosses of the arms of Danzig. However there is a record on the Bernstein database of a similar watermark of the crown with the two crosses. It is from Vilnius (Wilna), dated 1547 as below:

www.memoryofpaper.eu Piccard Online, Germany
http://www.piccard-online.de/?nr=126165

Although this Wilna 1547 watermark and the three from Belarus shown above all have the Danzig coat of arms, it is immediately apparent from the spacing of the vertical chain lines that the Belarus and Wilna papers are from different moulds or sources.

In addition to the helpful notes from Alexander Khmelevsky above, I am indebted to Dr Fred Hocker, Head of Research, Vasamuseet Stockholm for the following information (from emails in December 2019) about crowns:

> The Danzig/Gdańsk[35] coat of arms uses an open crown, which is the heraldic regalia of a duke rather than king, because Danzig was part of the Polish fief of the Dutchy of Prussia (later known as East Prussia), rather than Royal Prussia to the west. The cannon was cast for the city, and so carries its arms. Russian families who claimed a connection to Danzig as well as royal blood, such as the Sheremetovs noted above, substituted a closed crown, the regalia of a king, which the city never used.

However, open crowns were rare among watermarks, especially in comparison to the papal style tall crown, as may be seen from the following image which shows a sample of 1547 crown watermarks:

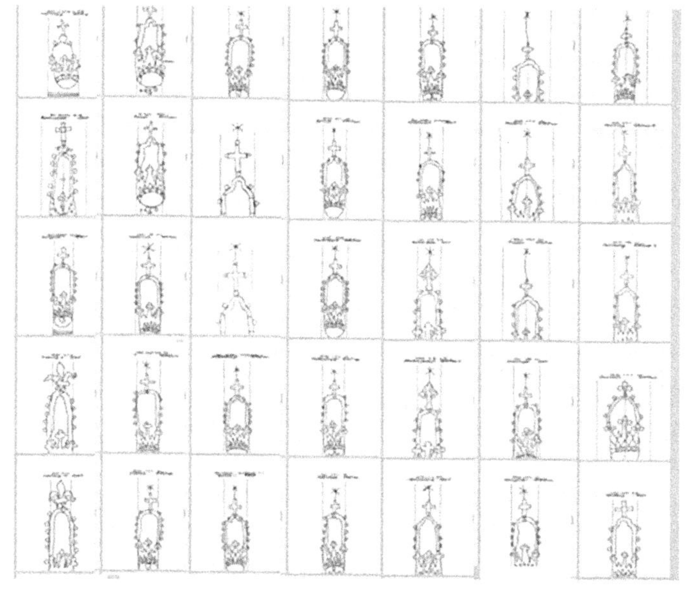

www.memoryofpaper.eu

Fourth Part

Case Study—Estonia—Number of Pages

The recommended procedure for watermark research calls for both front-lit and back-lit images to be archived. The following case study shows how either the front-lit or the back-lit images can be used to establish the number of pages even from the few remaining fragments of a rare book. Firstly, this case study uses the front-lit images to establish the number of pages in the only surviving fragments of the first book printed in Estonian as shown below. Secondly, the back-lit images are used to suggest that the type of paper used in that book was also used by Martin Luther for a personal letter.

The book is *Wanradt-Koell'i Katekismus,* 1535,[36] and it is the first printed in Estonian. Shortly after printing it was ordered that all copies, as described below, be destroyed. It was not until 1929 that fragments of it were discovered, in Tallinn, as part of a book's binding. The book is rightly regarded as emblematic of the enduring national spirit. For instance Paul Johansen and Hellmuth Weiss, *(Esimene eesti raamat anno 1535, Wanradt-Koell'i katekismus 1535 aastal,* Kultuur, 1956), Introduction, Olev Parlo, p. 7, claim, that, in the light of the severe and turbulent history of the country, the very survival of the fragments is a symbol of the nation. 'Kas see ei ole sümboolne ka kogu eesti rahva kohta, . . .' (Is this not symbolic for the entire Estonian people?).

The history of the book and the discovery of the fragments are indeed remarkable. Until 1929 it was believed that the oldest printed Estonian book was the 1632 *Hand- und Hausbuch* by Heinrich Stahl. However in November 1929 Dr. H. Weiss made a discovery in the binding of a book in the *Eestimaa Kirjanduse Ühing.* The binding is dated between 1541 and 1550. See R. Antik, (*Eesti raamat 1535-1935. Arengulooline Ülevaade / Arvulised Kokkuvõtted / Reproduktsioonid. Das Estnische Buch).* The book was made up of items printed in Wittenberg, Cologne and Strasbourg dated between 1530 and 1541. In the binding Dr. Weiss discovered fragments which proved to be from the above-mentioned Catechism. It is printed in two languages. Low German is on the left and Estonian is on the right. The Estonian also has marginal comments which are mostly corrections or linguistic alternatives. The book concludes with brief Low German notes on the Estonian language. From the few existing fragments it is clear that the contents follow Luther's Catechism—articles of faith, followed by the Lord's prayer, the Sacrament of Baptism and the Sacrament of the Altar.

The Commandments are absent—no doubt due to the fragmentary nature of the surviving pieces. Antik comments that the book is not a simple translation of Luther's Catechism but witnesses to being an independent work.

The author of the book was Simon Wanradt, Pastor at the St Nicholas Church in Tallinn. The translator was Johann Koell, Pastor at the Church of the Holy Spirit in Tallinn. No sooner had the book been printed than the Tallinn Council ordered that all copies be destroyed. The stated reason was that it contained not a few errors. Wanradt was in Tartu in 1529 and then moved to Wittenberg where he received his Master's degree before returning to Tallinn with a visit to Lübeck in 1532 apparently with the intent of recruiting preachers for Tallinn. A dramatic change happened in late 1536 / early 1537, just one year after the appearance of the *Katekismus*. Whatever the real cause of the change may have been it is recorded that Wanradt abandoned his wife and left with a disreputable Tallinn woman. He continued his life as a pastor dying in Danzig in 1563. Koell's life, in contrast to Wanradt's, was less turbulent. He remained in Tallinn. When the reformer chaplain Zacharias Hasse of the Holy Spirit Church died from 'flu in 1531 Koell took his place. His translation style is regarded as simple having an appeal to the masses. He died in 1540.

One of the fragments does allow the date (25 August 1535), name of printer (Hans Luft) and place of printing (Wittenberg) to be certain as may be seen in the following image:

drucket tho Wit
mberch dorch
ans Lufft/am
.XV. tage des
Mantes Au
M. D. XXXv.

Courtesy Tallinn City Archives

The writer was granted access to the few remaining fragments of *Katekismus,* 1535 (Simon Wanradt and Johannes Koell) for imaging in December 2015. One of the results was the digital reconstruction, as far as possible, of the 9 whole sheets of the octavo. Although the fragmentary nature of the material did not allow this, a partial reconstruction of the final gathering (I) was also achieved.

The book is an octavo. The existing fragments show signatures D, Dii, G, Giiij and I. There have been differing estimates as to the number of pages—in 1956 Johansen and Weiss suggested 'about 120',[37] though the same authors elsewhere suggest the number is in the region of 140.[38] However the evidence from the digital reconstruction of the final sheet I shows that those final pages, instead of being the eight leaves (16 pages) of a conventional octavo, are four leaves (eight pages). Therefore, as explained fully below, the total number of pages is 136.

Gatherings:

A, B, C, D, E, F, G, H;	8 x 8 leaves = 64 leaves (or 8 x 16 pages = 128 pages)
I:	1 x 4 leaves = 4 leaves (or 1 x 8 pages = 8 pages)
Total	136 pages.

The total number of pages is therefore 8 x 16 + 1 x 8 = 136.
The following images show this.

The Final Gathering 'I'

One side of the Final Gathering (the inner forme) is as below, with page numbers:

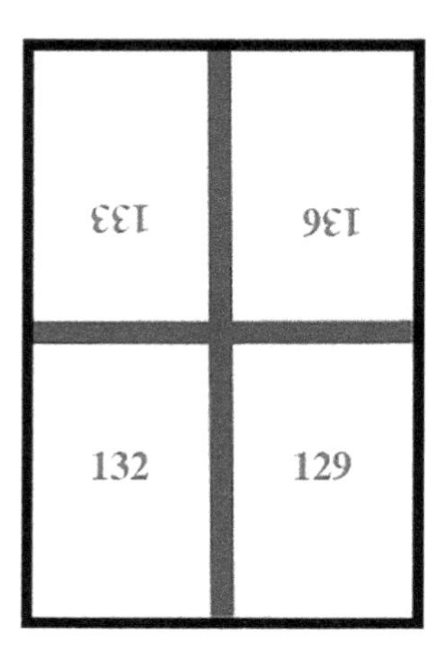

The other side of the Final Gathering (the outer forme) is as below, with page numbers:

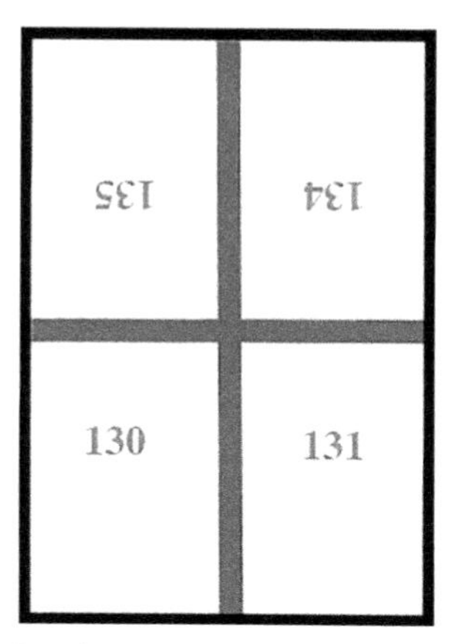

The final page of the book—136—a verso, is blank as below:

Courtesy Tallinn City Archives

This page (136) was digitally rotated and digitally joined with page 129 as shown below:

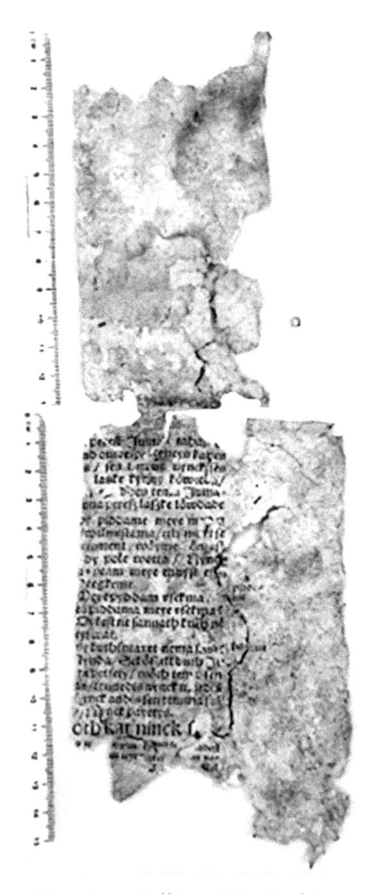

Courtesy Tallinn City Archives

The following image is a close-up of the 'join':

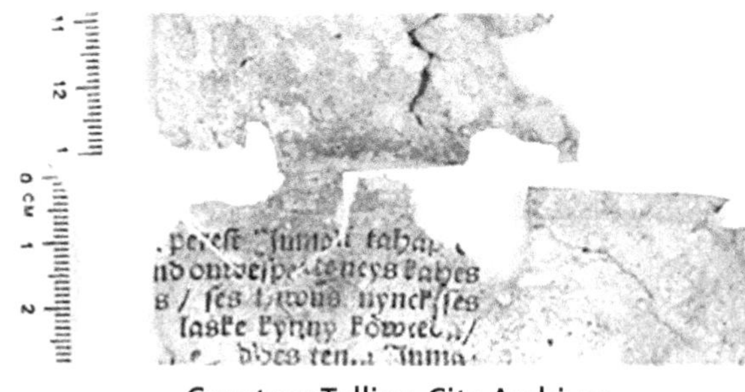

Courtesy Tallinn City Archives

When the same process is followed with pages 130 and 135 there are similarly impressive results. This is shown in the close-up below:

Courtesy Tallinn City Archives

It is therefore without doubt that there were 136 pages in *Katekismus* X.2615. The following image is the fullest possible reconstruction of the final gathering shown from both sides:

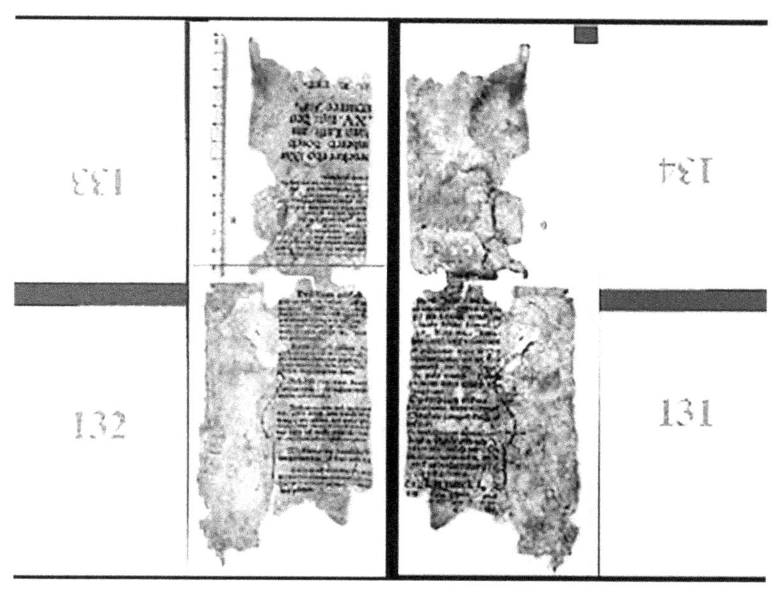

Courtesy Tallinn City Archives

Watermark in Simon Wanradt and Johannes Koell's, Katekismus, Wittenberg, Germany, 1535. Tallinna Linnaarhiiv X.2615

Since the book is an octavo, if any watermarks are present they are most likely to be found at the top of pages, but likely to be divided over more than one page. The image below, by way of example, shows the customary position of a watermark in a sheet of paper in a printed octavo:

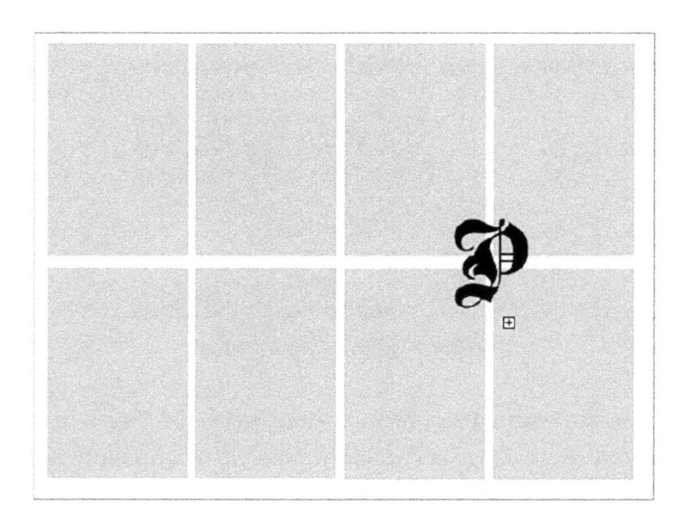

The eight rectos and the eight versos were all imaged using:

1. front lighting and
2. back lighting from a 1 mm thick electroluminescent light sheet.

It appears that all the fragments were from unbound sheets.

Happily one of the fragments (04) is of the top of two pages. The following images show 04 recto by front lighting and by back lighting, with the arrow showing the expected watermark position:

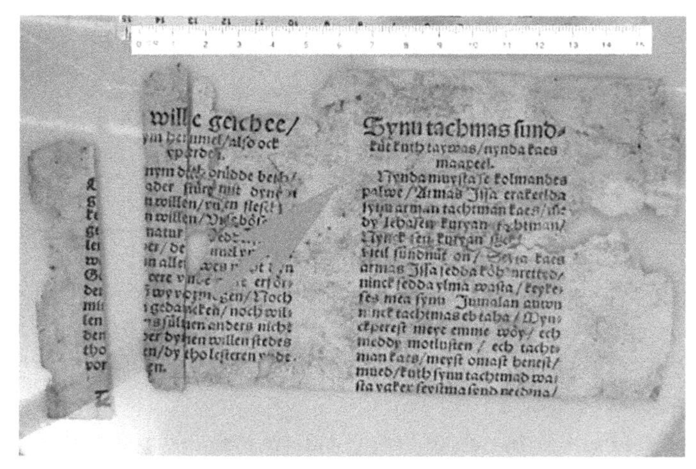

Courtesy Tallinn City Archives 04 Recto—front lighting

Courtesy Tallinn City Archives 04 Recto—Backlighting

After processing, it is now suggested that the very base two tips of a Gothic letter P are visible:

Courtesy Tallinn City Archives

The tips of the letter 'P' can be seen as dark curves above the '0' and the '2'. It is to be noted that the distance between the two tips is 1.8 cm.

The Search

Access to the Bernstein Memory of Paper allowed a search to be made under 'Witten(berg)' for the years 1530-1540. There are 8 of the distinctive tall crowns, animals, crests and one Gothic P (Bernstein Gothic P hereafter). It is now suggested that Bernstein Gothic P might be the same watermark as at 04 of X.2615.

Here is Bernstein Gothic P with a grid overlaid from which it can be seen that the distance between the tips is 1.8 cm.[39]

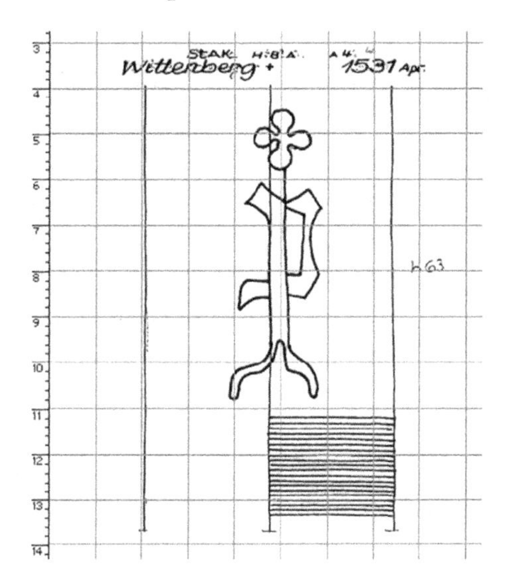

Here is a transcript of that letter by Martin Luther, 24 August 1531:[40]

— 252 —

376.

An Albrecht, Herzog von Preußen, v. 24. August 1531.

Aus Fabers Briefsammlung S. 6. De W. IV. 290.

Gnad und Friede in Christo. Durchleuchtiger, Hochgeborner Furst, gnädiger Herr! Wie E. F. G. an uns geschrieben und begehrt die Apologia oder Verantwortunge zu besehen: also haben wir gethan, und befinden auch, daß viel guter Grund und Ursachen gnugsam drinnen sind gestellet, und haben auf Ansinnen unsers gnädigsten Herrn des Kurfursten unser Meinung gen Hofe geschickt, und versehen uns, es sei E. F. G. numals zukommen, oder werde itzt mit Doctor Basilio kommen. Es hat uns auch wohl fur gut angesehen, daß nicht Noth sein solle, in angezeigten Stucken so gnau und weitläuftig sich herausgeben, weil aller Widersacher Art und Natur ist, wo sie den rechten Häuptgrunden nichts anhaben mugen, zwacken sie etwa ein Wort, und klügeln druber, damit die Sache aus der Bahn, und die Häuptgrunde aus den Augen kommen und den Schein verlieren, wie mir bisher täglich geschehen ist in alle meinem Schreiben: darumb mit solchen Leuten zu handeln, das Beste ist, kurz und feste hindurch, und nicht sich von den Häuptgrunden fuhren lassen. Doch wird E. F. G. solchs alles wohl besser wissen zu verstehen, dann wir anzeigen konnen. E. F. G. sei nur getrost und lasse sich solchs nicht bekommern, Gott wirds wohl machen. Hätten sie nicht hievon zu plaudern, so mußten sie ein anders haben, so haben sie so mehr dieß, als ein anders; dem Teufel kann Niemand sein Maul stopfen, er muß plaudern. Christus unser Herr stärke und troste E. F. G. zu thun und zu leiden allen seinen gnädigen Willen, Amen. 24. Augusti, 1531.

E. F. G.

williger

Martinus LutheR

von wegen unser aller.

Here is an image of the letter:

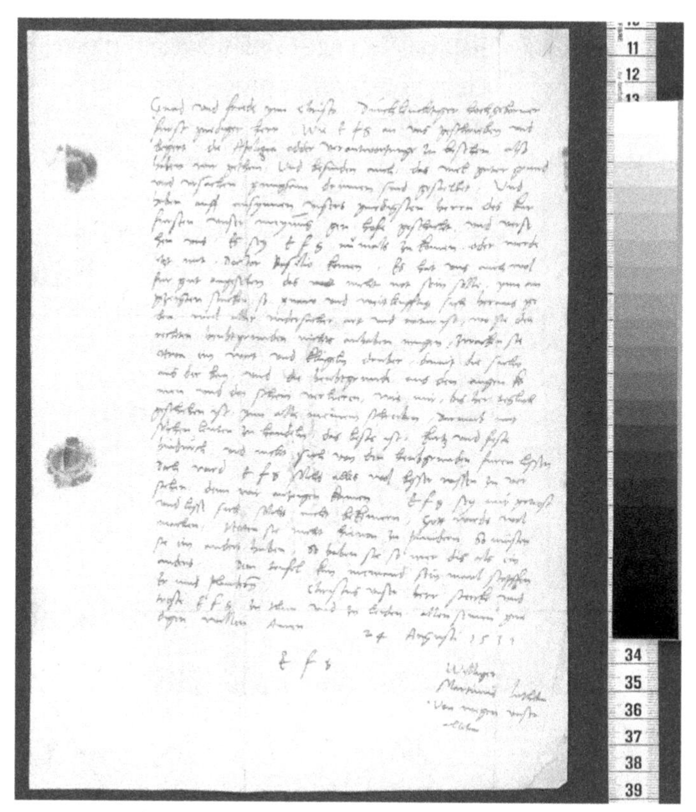

Courtesy of Geheimes Staatsarchiv Preußischer Kulturbesitz

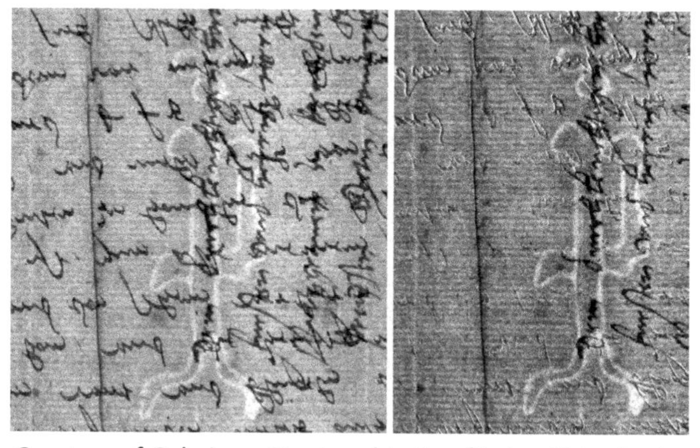

Courtesy of Geheimes Staatsarchiv Preußischer Kulturbesitz

The back-lit image on the left is from the original image kindly supplied by the Geheimes Staatsarchiv Preußischer Kulturbesitz. The image on the right is the result of processing that back-lit image with the front-lit image. The digital removal of overprinting is impressive. The process has been fully described above under Image Processing and Archiving.

This Case Study has shown that either front-lit or back-lit images can be used, in certain cases, to establish the number of pages in a rare book, even when pages are missing. The Case Study also shows the potential of watermark research for revealing unexpected cultural and historical connections. In the case cited the book—*Wanradt-Koell'i Katekismus*, 1535—had been printed in Wittenberg. The evidence from watermark research as given above indicates that the paper in that book appears to have a Gothic P watermark similar to the paper used by Martin Luther of Wittenberg in 1531 for a letter on behalf of Doctor Basilio to Duke Albrecht.

Fifth Part

Composition and dating

The end results of combining front-lit and back-lit images of multiple pages into a single multi-layered file are that:

1. security is enhanced by the *PaperPrint* procedure, as described above
2. a flexible research resource is created—as shown above and by the following example concerning the composition and dating of the first book printed in Finnish.

Here is an image of the single Photoshop file of the Inner Forme of Gathering A of the Uppsala University copy of that first book, which is an octavo. It is *Abckiria Michael Agricola Christiano salutem,* by Mikael Agricola (ca. 1510-1557). The image below shows the whole sheet, as originally printed, by front lighting. The file consists of 17 layers. There are 8 correctly arranged images, which show 8 pages front lit.

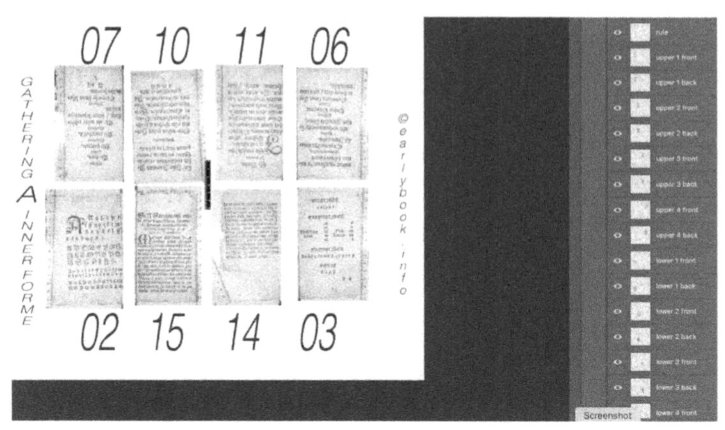

Courtesy of Uppsala University Library Rar. 10:233 1549

There are also 8 correctly arranged underlying back-lit images, which show those eight pages back lit. By selecting one by one the 8 front-lit images and by 'turning them off'/deselecting them, all 8 back-lit images are revealed, including the watermark. In the following image the 8 front-lit images have been 'turned off'/deselected, as indicated in the right-hand column:

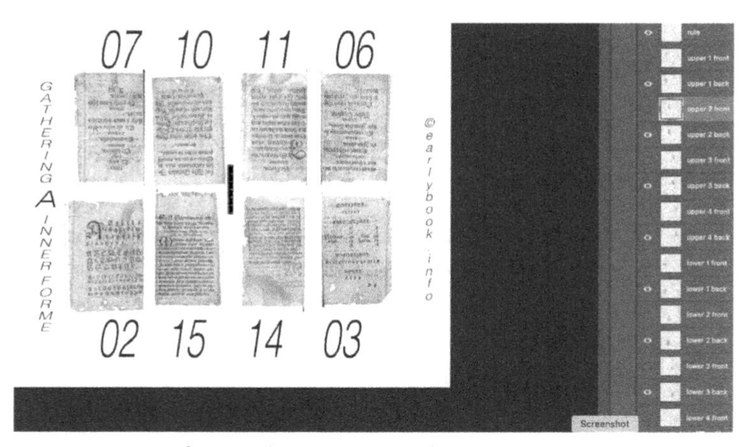

Courtesy of Uppsala University Library Rar. 10:233 1549

In addition, selection of the Rule layer allows the mm rule to be moved and rotated so that accurate measurements may be made. This large Photoshop file is downloadable from the Downloads page.

Firstly, the effectiveness of this file for examining the composition or make up of the book is easily shown: even a casual examination of pages 15 and 10 of either of the images above shows that the uppermost line of text in both images is the same. The following images show the uppermost lines of pages 15 and 10:

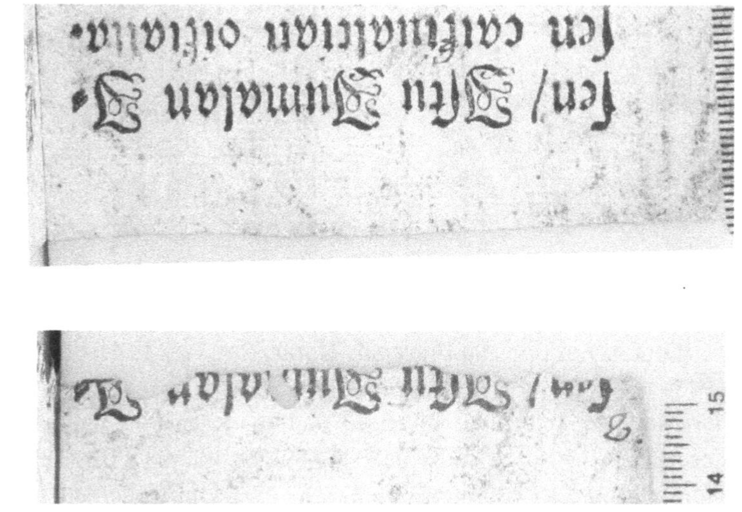

Courtesy of Uppsala University Library Rar. 10:233 1549 Front-Lit

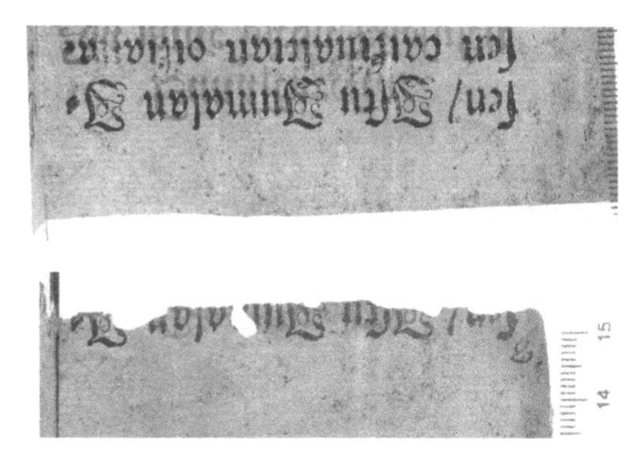

Courtesy of Uppsala University Library Rar. 10:233 1549 Back-Lit

The misaligned (vertical) chain lines above clearly show that different sheets of paper were used, and the line of text which appears at the top of both images proves that different sheets of paper were used. This means that pages 15 and 10 cannot have come from the same sheet.

Secondly, this file is effective for querying the currently accepted date of the book printed by Amund Laurentsson in Stockholm. Sweden still holds two complete copies and one incomplete (earlier) copy of this rare book. There are also two in Helsinki. According to the Finnish National Library, the date of their copy is 1543.[41] According to the Uppsala University Library catalogue, their copy is to be dated 1549.[42] This uncertainty about the date has been acknowledged by Laitinen and Schoolfield, who wrote 'The exact publication date of Agricola's earliest work, the *ABCKiria*, is uncertain; apparently, it lay between 1537 and 1543, and the last of probably three editions came out in 1559'.[43] The same highly distinctive watermark, as shown below, of a sphere with a star was found in copies of *ABCKiria* at Uppsala University and at the Finnish National Library.

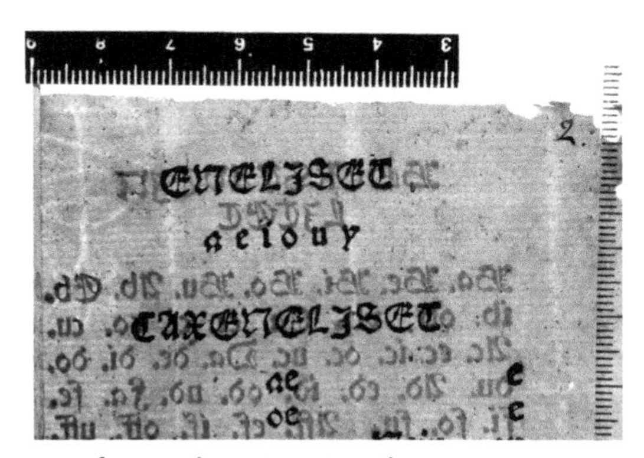

Courtesy of Uppsala University Library Rar. 10:233 1549

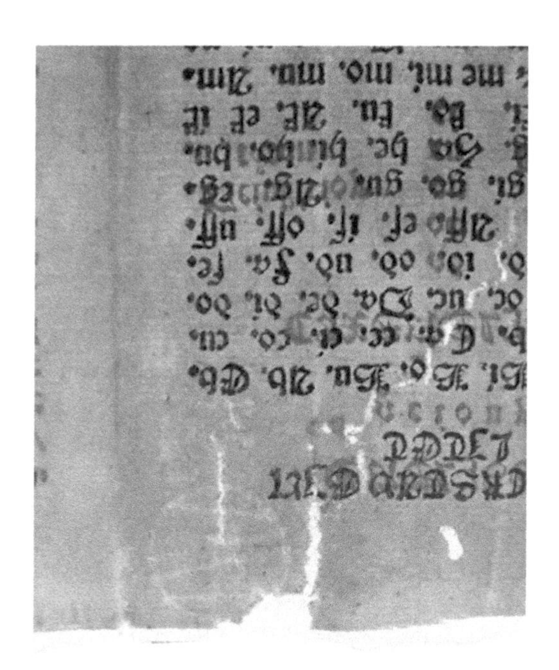

ABCKiria, aii, Helsinki National Library

The same watermark is at Zaragossa, dated 1553:

Zaragossa, 1553 DPZ-0633.

The on-line Bernstein Memory of Paper database was searched. Motifs were 'Estrella' (star), 'Orbe' (Orb), Sphere and Kugel. The most significant result was that there are no recorded matching watermarks before 1550. Likhachev has 15 relevant matches for Sphere (France and Russia) for 1550-1559.[44] Briquet has 24 relevant matches for Sphere (23 France, 1 in Vicenza, Italy) for 1550-1559. He states that there is no doubt that the paper is from the French south-west (Angoulême) region.[45]

The Bernstein datafile has allowed the growing distribution of that watermark to be mapped. The first recorded is Bordeaux, 1550. By 1553 it had reached Agen and Zaragossa. Later it is recorded in Rostock, reaching into Russia as shown below:

It is concluded that there are good grounds for questioning the date of 1543 given for the Helsinki copy and the date of 1549 given for the Uppsala copy of *ABCKiria*. Thanks to watermark research it can be deduced that a date in the 1550s is much more likely.

Downloads

1 Notes

Blank pdf of Notes is downloadable from here:
https://www.dropbox.com/s/sogjkq3od1uqcjy/notes.psd?dl=0

---

2 Octavo Assemble

Movie showing how an octavo is assembled from one sheet of paper.
https://www.dropbox.com/s/y72l8fccikmr83b/OCTAVOmake.mp4?dl=0

---

3 Octavo Both Sides

Download from here the pairs of images which can be printed back to back thereby recreating the sheet Gathering A, as originally printed from the Vilnius copy of the first book printed in Lithuanian. Both pairs are available below in Large and Smaller sizes. The book is Martynas Mažvydas, *CATECHISMVSA PRAsty Szadei,* Weinreich, Jan. (Kaliningrad, 1547).

Outer Forme 481 mb 528 cm wide Large
https://www.dropbox.com/s/ll2413xy4zc1fk1/A%20O.jpg?dl=0

Outer Forme 3.5 mb 45 cm wide Smaller
https://www.dropbox.com/s/6yfda15tsj59r0k/A%20O%2045cm.jpg?dl=0

Inner Forme 497 mb 540 cm Large
https://www.dropbox.com/s/7yru49r25qqinoo/A%20I.jpg?dl=0

Inner Forme 3.45 mb 45 cm Smaller
https://www.dropbox.com/s/yh25hkeu6mehguc/61%20D.jpg?dl=0

4 *ABCKiria* 17 Layers Gathering A. 290 mb. 372 cm wide.

https://www.dropbox.com/s/5bbleddxkganuw8/62.psd?dl=0

Endnotes

1 Michael Blanding, *The Map Thief: The Gripping Story of an Esteemed Rare-Map Dealer Who Made Millions Stealing Priceless Maps*, New York: Gotham Books, 2014.

2 Miles Harvey, *The island of lost maps: a true story of cartographic crime*, New York: Random House, 2000.

3 Michael Vinson, *Bluffing Texas Style—The Arsons, Forgeries, and High-Stakes Poker Capers of Rare Book Dealer Johnny Jenkins*, Norman: University of Oklahoma Press, 2020.

4 Michael Blanding, *The Map Thief: The Gripping Story of an Esteemed Rare-Map Dealer Who Made Millions Stealing Priceless Maps*, New York: Gotham Books, 2014, p. 112.

5 https://www.lexpress.fr/actualite/societe/un-ancien-de-la-bnf-en-prison_462551.html.

6 In 2020 Gregory Priore, the former archivist of the Carnegie Library's rare book room, and John Schulman, owner of Caliban Book Shop in Pittsburgh, pleaded guilty to theft and receiving stolen property. The stolen items were valued at $8,000,000. Greg Priore was sentenced to three years' house arrest and 12 years' probation, and Schulman to four years' house arrest and 12 years' probation.

7 The Royal Horticultural Society's Lindley Library: Safeguarding Britain's Horticultural Heritage. 22 May 2013.

8 [Silvester Kosov], Treatise on the Sacraments, Kiev, Monastery of the Caves, 1657. Marsh G.3.4.27. These two copies are bound in the western style in one volume together with the 1656 sermon on the plague by Patriarch Nikon. See Cleminson, Cyrillic books printed before 1701 in British and Irish collections: a union catalogue / compiled by Ralph Cleminson . . . [et al.], British Library, 2000.

9 The etching, held at the Rembrandthuis in Amsterdam, is of Saint Jerome by a pollard willow. https://blogs.bl.uk/european/2014/02/light-links-lithuania-to-k%C3%B6nigsberg-kiev-to-rembrandt.html.

10 https://www.sothebys.com/en/videos/the-bay-psalm-book-americas-first-printed-book Accessed 1st August 2020.

11 Richard L. Hills, *A Technical Revolution in Papermaking, 1250-1350,* art. in 'Looking at paper: evidence & interpretation'. Ottawa, Canadian Conservation Institute, 2001, 105-111.

12 Alfred H Shorter, ed. Richard L. Hills, *Studies on the History of Papermaking in Britain,* Variorum, 1993, and Alfred H. Shorter, *Paper Mills and Paper Makers in England. 1495-1800,* The Paper Publications Society, 1957. Shorter describes those early English mills as follows:

> John Tate (born about 1448) is credited with being the first person to produce paper in England. His Hertford mill not only enjoyed royal attention but the watermark was distinct and has been described as an eight-pointed star, the petals of a flower or the spikes of a wheel set within a double circle. It is a watermark unique to Tate for, surprisingly, it has been used nowhere else. It appears with slight differences owing to different moulds having been used. The watermark enables us to identify as Tate's paper the printed version of a Papal Bull expressing pleasure in the marriage of Henry Tudor and Elizabeth of

York, and also recognising Henry as the rightful occupant of the English throne. Richard L. Hills, *Papermaking in Britain, 1488-1988*, Athlone, London, 1988.

13 There is a fine selection of variant mould types at Simon Barcham Green https://papermoulds.typepad.com/

The URL above and all below were accessed 4th August 2020. Three examples are cited below to show variant types of moulds:

1. Apparently the mid-Wales Gwas Gregynog (Gregynog Press) double mould had the chain lines spaced irregularly in order to give the impression of age. https://papermoulds.typepad.com/photos/m263_gwas_gregynog_laid/m263b-watermark.html.

2. A mould could be made with a central tear wire allowing one sheet of paper to be torn so that two smaller sheets were produced as from the 1954 mould for the Queen's bankers Coutts & Co. https://papermoulds.typepad.com/photos/m260_coutts_co_bankers_lo/p1020399.html.

3. Even a two-sheet mould might have more than one watermark, such as the 1894 Edwin Amies mould for security printers Waterlow & Sons Ltd. Each side has Waterlow's name and an ornate composite watermark of a snake with its tail in its mouth (an ouroboros) and the Etruscan/Roman fasces.
https://papermoulds.typepad.com/photos/m_207_waterlow_sons_ltd_l/m-207-c-waterlow-sons-ltd-london-ouroboros.html.

14 Ronald B. McKerrow, *An Introduction to Bibliography for Literary Students,* St Paul's Bibliographies, Winchester, Oak Knoll Press, 1994, 170-171.

15 Collation of *The Grete herbal,* 615.3 GRE, kindly supplied by Ms. Debbie Lane, formerly of the Lindley Library, RHS: 2^{O}. $+^{6}$ $(\pm+^{1\text{-}6})$ A-K^{6} L^{6} $(\pm L^{6})$ M^{6} N^{6} (±N2) O-Y^{6} Z^{6} (±Z4) Aa-Ee^{6} [$3 signed (Uiii as Uiiii); leaves +1-6 and Z4 are facsimile reproductions; woodcuts and some text at the tops of leaves $Y2^{a\text{-}b}$, $Y6^{a\text{-}b}$ and $Bb2^{a\text{-}b}$ are hand-drawn facsimiles.

16 C. M. Briquet, *Les Filigranes,* Paper Publications Society, 1968.

17 Available from: https://www.gimp.org/downloads/ Accessed 13th October 2018.

18 http://www.soane.org/

19 http://sites.scran.ac.uk/ada/documents/adam_r.htm

20 Johannes Potken, *The Psalms, followed by Sacred Canticles and the Song of Songs.* Rome, 1513. King's College London, Foyle Special Collections [Marsden Coll.] R8/1.

21 *CATECHISMVSA PRAsty Szadei: Makslas skaitima raschta yr giesmes del kriksczianistes bei del berneliu jaunu nauiey sugulditas.* KARALIAVCZVI, Weinreich, Jan. Kaliningrad/Königsberg, 1547.

22 University Library in Toruń, Nicolaus Copernicus University Library. https://szukaj.bu.umk.pl/discovery/search?query=any,contains,Pol.6.II.189%20adl.3&tab=Everything&search_scope=MyInst_and_CI&vid=48OMNIS_UMKWT:UMK&offset=0

23 Vilnius University 000408215.

24 Sefarewicz, Jan, art. 'Un acrostiche de Mažvydas' (An Acrostic for Mažvydas) (*Prace Filologiczne,* XVIII, 1938), 7-8.

25 Thanks to a generous grant from the Lutheran Church Missouri Synod the author was able to record the entire musical content of the Holy Hymns. The recordings are of the Commandments, the request for the Holy Spirit, the Lord's Prayer, Psalm 103, Psalm 51, the Creed, Dies est lætitiæ, the Sacrament of the Altar and Christe qui lux es et dies. They are all available online from:
https://www.dropbox.com/sh/lwswuhkupx9ksup/AAA_USA8l8li346N-B6FFXhva?dl=0

The recordings were made at the Lithuanian Roman Catholic Church of St Casmir in London thanks to the Rev. Petras Tverijonas. The music was prepared and performed by Ms. Aušrinė Aurelia Apanavičiūtė. Here is a decode about the file names:
m = music
w = words
mw = music and words
The numbers below refer to the pages of the Vilnius book.

40, 41, 42	Commandments
43, 46, 47	request Holy Spirit
48	Lord's Prayer
59	Psalm 103
64	Psalm 51
68	Creed
72	Dies est lætitiæ
75	Sacrament of the Altar
77	Christe qui lux es et dies

26 Back-lit images of every page of the Vilnius copy of the book were archived. Backlighting was provided by a 1 mm thick electroluminescent sheet placed under the paper. http://www.earlybook.info shows the Early Book Imaging System.

27 http://www.memoryofpaper.eu/BernsteinPortal/appl_start.disp. Accessed 23rd May 2017.

28 Search = /insignia / crown / without arch / detached, without additional motif / vertical / *Mittelzinken* consisting in two lines / extremity leaf-/lily-shaped / *Reif einteilig* / without jewellry. Watermark DE4620-PO-50936.

29 Gordon B. Ford, Jr., *The Old Lithuanian Catechism of Martynas Mažvydas (1547)*, Assen, Van Gorcum, 1971, XII.

30 Richard Frost, *The Oxford History of Poland-Lithuania. Vol I: The Making of the Polish-Lithuanian Union, 1385-1569*, Oxford, Oxford University Press, 2015, 394.

31 Full details are at: BOBORYKIN: https://gerbovnik.ru/arms/619.html.

KOLYCHEV: https://gerbovnik.ru/arms/177.html.

Counts KONOVNITSYN: https://gerbovnik.ru/arms/1417.html.

LODYGIN: https://gerbovnik.ru/arms/1263.html.

NEPLYUEV: https://gerbovnik.ru/arms/759.html.

Counts SHEREMETEV: https://gerbovnik.ru/arms/160.html.

YAKOVLEV: https://gerbovnik.ru/arms/178.html.

32 Zoya Vasil'evna Uchastkina, *History of Russian Hand Paper-mills and their watermarks . . . Edited and adapted for publication in English by J. S. G. Simmons [with the assistance of B. J. van Ginneken-van de Kasteele]. [Monumenta chartæ papyraceæ historiam illustrantia. no. 9.]*, Paper Publications Society: Hilversum, 1962, Map III. Each square division marks 75 km (47 miles) approximately.

33 Zoya Vasil'evna Uchastkina, *History of Russian Hand Paper-mills and their watermarks . . . Edited and adapted for publication in English by J. S. G. Simmons [with the assistance of B. J. van Ginneken-van de Kasteele]. [Monumenta chartæ papyraceæ historiam illustrantia. no. 9.]*, Paper Publications Society: Hilversum, 1962, Plates 364 and 365. Page 792.

34 I would like to record my thanks to Alexander Khmelevsky, co-moderator of site gerbovnik.ru, for explaining that the second letter 'C' does not mean 'СВЕТЛОСТЬ' but 'СИЯТЕЛЬСТВО'. There are different forms of address for titled persons in Russia—'СВЕТЛОСТЬ' is only for the Serene Princes, but 'СИЯТЕЛЬСТВО' is the correct form for counts and princes. Although the English GRACE is more consistent with the title 'СИЯТЕЛЬСТВО', the Russian 'СВЕТЛОСТЬ' corresponds to the English SERENE HIGHNESS. Translation of the Russian 'ГРАФ' into English is only 'COUNT'. The English DUKE is Russian 'ГЕРЦОГ', which is comparable to the German GRAF and HERZOG.

35 Danzig was a prominent city from the later Middle Ages onward, with widespread influence and contacts. The city long claimed a largely independent status, and in fact was an independent city-state (the Free City of Danzig) between the World Wars, as one of the terms of the treaties which ended World War I.

36 X.2615, Tallinna Linnaarhiiv.

37 Paul Johansen and Hellmuth Weiss *Esimene eesti raamat anno 1535, Wanradt-Koell'i katekismus 1535 aastal*, Kultuur, 1956, p. 9. 'Seda aluseks võttes võis raamatul olla umbes 120 lehekülge'. (On this basis the book has about 120 pages). This 1956 estimate follows the 1936 estimate by R. Antik who claimed that 'Oktaavkaustas trükitud raamat on sisaldanud umbes 120 lehekülge. . . . Das in Oktavformat gedruckte Buch hat ungefähr 120 Seiten enthalten' (The Octavo format printed book consists of some 120 pages). *Eesti raamat 1535-1935. Arenguloonline Ülevaade / Arvulised Kokkuvõtted / Reproduktsioonid. Das Estnische Buch Entwicklungsgeschichtlicher Überblick / Zahlenmässige Nachweise / Abbildungen1535-1935.* Tartu, 1936, p. 22.

38 Paul Johansen and Hellmuth Weiss, Art. Bruchstücke eines niederdeutsch-estnischen Katechismus vom Jahre 1535, in *Beiträge zur Kunde Estlands.* p. 96, 'Danach können wir auf einen Umfang von ca. 140 Seiten schließen'. (So we can conclude that there are about 140 pages.)

39 Here are details from the Bernstein site: DE4620-PO-110662 http://www.wasserzeichen-online.de/wzis/?ref=DE4620-PO-110662.

Showing, in this order (own translation from the German): letters / digits; letter P; free, gothic shape, with markings; flower/leaf; four-leaved, without any additional mark, without a horizontal line; arch end behind the shaft; Split shaft end without frills; bow end without thorn; leaves round, without a stamp. Königsberg, H.B.A. A 4 1531, Wittenberg Aussteller: Martin Luther Papier P, Kleinformat Breite 20 mm, Höhe 62 mm. http://www.piccard-online.de/?nr=110662.

40 *Dr. Martin Luther's sämmtliche Werke* (Bd. 1-20, herausgegeben von J. G. Plochmann; Bd. 21-65, bearbeitet von J. K. Irmischer; Alphabetisches Sach-Register. . . , herausgegeben von J. K. Irmischer) Erlangen, 1826-57.

41 https://kansalliskirjasto.finna.fi/Record/fikka.3885317
Accessed 21 April 2021.

42 http://www.alvin-portal.org/alvin/attachment/document/alvin-record:104738/ATTACHMENT-0030.pd
f Accessed 21 April 2021.

43 Laitinen, Kai and Schoolfield, George C., *A History of Finland's Literature.* vol 4. Histories of Scandinavian Literature. Ed. Schoolfield, George C., tr. Binham, Philip, University of Nebraska Press, 1998, p. 35-36.

44 Nikolaĭ Petrovich Likhachev, *Likhachev's watermarks: an English-language version,* ed. J. S. G. Simmons and Bé van Ginneken-Van de Kasteele. Amsterdam: Paper Publications Society, 1994.

45 Charles M. Briquet, *Les Filigranes.* Paper Publications Society, 1968, vol. 4, p. 689.

Bibliography

Agricola, Mikael. *ABCKiria.* [Amûd Lauritzen poijalda] [Herran syndimen wodesta, 1559].

________. *Suomenkielisen kirjallisuuden esikoinen. Michael Agricolan Abckiria. Alkuperäisessä muodossa uudestaan painattanut K. G. Leinberg.* [A facsimile of the edition published by A. Laurentsson at Stockholm in 1549.] Jywäskylässä, 1864.

________. *Abckiria.* Helsinki: Suomalaisen Kirjallisuuden Seura, 1982.

________. and Kaisa Häkkinen. *Mikael Agricola: Abckiria: kriittinen editio/toimittanut.* Helsinki: Suomalaisen Kirjallisuuden Seura, 2007.

Ainsworth, Henry. *The book of Psalmes: Englished both in Prose and Metre. With annotations, opening the words and sentences, by conference with other scriptures.* Amsterdam, 1612.

Antik, R. *Eesti raamat 1535-1935: Arengulooline* Ülevaade, *Arvulised Kokkuvõtted, Reproduktsioonid: Das estnische Buch entwicklungsgeschichtlicher* Überblick, *zahlenmäßige Nachweise, Abbildungen 1535-1935.* Tartu, 1936.

Bogdanov, Andrey Petrovich. Основы филиграноведения: история, теория, практика. [Bases of Watermarks Study: History, Theory, Practice]. Moscow, 1999.

Breydenbach, Bernhard von. *Peregrinatio in terram sanctam.* Mainz, 1486.

Briquet, Charles-Moïse. *Les Filigranes. Dictionnaire historique des marques du papier dès leur apparition vers 1282 jusqu'en 1600: A facsimile of the 1907 edition with supplementary material contributed by a number of scholars,* edited by Allan Stevenson. Amsterdam: Paper Publications Society, 1968.

Calvin, Jean. *Harmoniæ ex evangelistis Mattheo, Marco et Luca.* Avignon, 1582.

Cisneros, Cardinal Francisco Ximenes de. *Complutensian Polyglot.* Complutense University, Alcalá de Henares, 1520.

Crawford, Osbert Guy Stanhope. *Ethiopian itineraries, circa 1400-1524: including those collected by Alessandro Zorzi at Venice in the years 1519-24.* Cambridge: Published for the Hakluyt Society at the University Press, 1958.

Daniell, David D. *William Tyndale: A Biography.* New Haven: Yale University Press, 1994.

Daye, Stephen. *The Whole Booke of Psalmes Faithfully Translated into English Metre.* Cambridge: Stephen Daye, 1640.

Gaudriault, Raymond and Thérèse Gaudriault. *Filigranes et autres caractéristiques des papiers fabriqués en France aux XVIIe et XVIIIe siècles.* Paris: CNRS Éditions, 1995.

Haraszti, Zoltán. *The Enigma of the Bay Psalm Book.* Chicago: University of Chicago Press, 1956.

Heawood, Edward. 'Transactions of the Bibliographical Society', *The Library.* London, 1930.

________. 'Sources of early English paper supply. London, Extracts from the Transactions of the Bibliographical Society'. *The Library*, 1931-1947.

Hills, Richard L. 'A Technical Revolution in Papermaking, 1250-1350', in *Looking at paper: evidence & interpretation, Toronto 1999*. Ottawa: Canadian Conservation Institute, 2001.

Hunter, Dard. *Papermaking: The History and Technique of an Ancient Craft: second edition, revised and enlarged.* London: Cresset Press, 1957.

Johansen, Paul and Hellmuth Weiss. *Esimene Eesti raamat, anno 1535: Wanradt-Koell'i katekismus 1535, aastal.* New York: Kultuur, 1956.

________. 'Bruchstücke eines niederdeutsch-estnischen Katechismus vom Jahre 1535' [Fragments of a Low German – Estonian 1535 Catechism]. *Beiträge zur Kunde Estlands* 15, no. 4 (1930): 1-40.

Kelly, Samantha. 'The Curious Case of Ethiopic Chaldean: Fraud, Philology, and Cultural (Mis) Understanding in European Conceptions of Ethiopia', *Renaissance Quarterly* 68, no. 4 (Winter 2015): 1127-1264.

Labarre, Émile Joseph. Gen. ed. *Monumenta chartæ papyraceæ historiam illustrantia.* vol. 1-vol. 15. Paper Publications Society, Hilversum. *Illustrated Collection of the History of Paper.*

Schoolfield, George C. *A History of Finland's Literature: Histories of Scandinavian Literature.* Translated by Philip Binham. Lincoln: University of Nebraska Press, 1998.

Lenz, Eduard von and Count Sergey Dmitrievich Scheremetew. *Die Waffensammlung des Grafen S. D. Scheremetew in St. Petersburg mit 26 Lichtdruck Tafeln.* Leipzig: Hiersemann, 1897. *The Armoury of Count S. D. Scheremetew in St. Petersburg with 26 Photographs.*

Luther, Martin. *Dr. Martin Luther's sämmtliche Werke.* (Bd. 1-20, herausgegeben von J. G. Plochmann; Bd. 21-65, bearbeitet von J. K. Irmischer; Alphabetisches Sach-Register, . . . herausgegeben von J. K. Irmischer) [*The Collected Works of Dr. Martin Luther.* (Volumes 1-20 published by J. G. Plochmann; Volumes 21-65 by J. K. Irmischer; Alphabetical Index . . . published by J. K. Irmischer.)], Erlangen 1826-57.

Marot, Clément and Théodore de Bèze. *Les Pseaumes de David, mis en rime francoise.* Anvers: Plantin, 1564. *The Psalms of David, put into French verse.*

Mažvydas, Martynas. *CATECHISMVSA PRAsty Szadei.* Weinreich-Königsberg-Kaliningrad: H. Veinreicho, 1547. Title page of Toruń copy at: http://www.epaveldas.lt/object/recordDescription/VUB/VUB01-000541149.

Mažvydas, Martynas and Gordon B. Ford Jr. *The Old Lithuanian Catechism of Martynas Mažvydas (1547).* ed. and trans. by Gordon B. Ford Jr. Assen: Van Gorcum, 1971.

Mosser, Daniel, Michael Saffle and Ernest W. Sullivan. *Puzzles in paper: Concepts in Historical Watermarks: essays from the International Conference on the History, Function, and Study of Watermarks in Roanoke, Virginia.* New Castle: Oak Knoll Press, 2000.

Pablos, Juan. *Breve y más compendiosa doctrina Christiana en lengua Mexicana y Castellana* [Brief and full account of Christian doctrine in Mexican and in Spanish]. Mexico City: Casa de Juan Cromberger, 1539.

Paper Publications Society. *A Short Guide to Books on Watermarks*. Hilversum: Paper Publications Society, 1955.

Pazdni͡akoŭ, Valeryĭ Si͡ami͡onavich. *Filihrani arkhiŭnykh dakumentaŭ Belarusi XVI - pachatku XX st. Watermarks of archival documents of Belarus of the 16th - early 20th century*. Minsk: BelNDIDAS, 2013.

Potken, Johannes. *The Psalms, followed by Sacred Canticles and the Song of Song*. Rome, 1513.

________. *Psalterium in quatuor linguis: Hebraea, Graeca, Chaldaea [i.e. Ethiopic], Latina: preceded by, Introductiunculae in tres linguas externas, Hebraea, Graeca, Chaldaea* [The Psalms in four languages: Hebrew, Greek, Chaldee (Ge'ez), Latin; preceded by an Introduction to the three foreign languages of Hebrew, Greek and Chaldee]. Edited by Joannes Potken and Joannes Soter. Cologne, 1518.

Salette, Arnaud de. *Los Psalmes de David metuts en rima bernesa, A Ortes. per Louïs Rabier* [The Psalms of David put into verse in the language of Béarn, At Orthez by Louis Rabier]. Orthez, 1983.

________. *Los Psalmes de David metvts en rima bernesa:* Édition *critique bilingue par Robert Darrigrand, sur le texte de l'édition publiée en 1583 par Louis Rabier*. [The Psalms of David put into verse in the language of Béarn: Bilingual Critical Edition by Robert Darrigand, from the text published in 1583 by Louis Rabier]. Paris: Champion, 2010.

Safarewicz, Jan. "Un acrostiche de Mažvydas". *Prace Filologiczne* 18, 1938.

Shorter, Alfred H. *Paper Mills and Paper Makers in England: 1495-1800*. The Paper Publications Society, 1957.

________ ed. Richard L. Hills, *Studies on the History of Papermaking in Britain*. Milton Park: Variorum, 1993.

Stahl, Heinrich. *Hand- und Hauszbuches Für die Pfarherren und Hauszväter Esthnischen Fürstenthumbs, Ander Theil, Darinnen das Gesangbuch, Zusampt den Collecten und Prefationen. In Teutscher und Esthnischer Sprache angefertiget, und auff eygenen Kosten zum Druck* übergeben [The Hand- and Homebook for Pastors and Fathers of Households in the Estonian Principality, including the Song Book, together with the Collects and Introductions. Presented in German and Estonian with both on the page]. Tallinn: Christoff Reusner, 1637.

Treveris, Peter. *The Grete Gerball whiche geueth parfyt knowledge and vnderstandyng of all maner of herbes [and] there gracyous vertues whiche god hath ordeyned for our prosperous welfare and helth, for they hele [and] cure all maner of dyseases and sekenesses that fall or mysfortune to all maner of creatoures of god created, practysed by many expert and wyse maysters, as Auicenna [and] other. [et]c. Also it geueth full parfyte vnderstandynge of the booke lately prentyd by me (Peter treueris) named the noble experiens of the vertuous handwarke of surgery*. Southwarke, 1526.

Tschudin, Walter Friedrich. *The Ancient paper-mills of Basle and their marks*. Hilversum, Paper Publications Society, 1958.

Tyndale, William. *The Newe Testame[n]t as it was written, and caused to be writte[n], by them which herde yt. To whom also oure saveoure Christ Jesus commaunded that they shulde preache it vnto al creatures*. Worms: Peter Schöffer, 1526.

________. *The obediēce of a Christen man and how Christē rulers ought to governe, where in also (yf thou marke diligently) thou shalt fynde eyes to perceave the crafty conveyaūce of all iugglers. Hans luft: at Marlborow in the lāde of Hesse.* Antwerp: J. Hoochstraten, 1528.

Uchastkina, Zoya Vasil'evna. *History of Russian Hand Paper-mills and their watermarks.* Edited and translated by J. S. G. Simmons. Hilversum: Paper Publications Society, 1962.

Ullendorff, Edward. *The Ethiopians: An Introduction to Country and People.* Stuttgart: Franz Steiner, 1990.

Urbain, Daniel. "La version tchèque (1587) du Psautier de Genève" [The Czech Version of the Geneva Psalter]. *Revue de littérature comparée* 3 (1988): 367-375.

Wanradt, Simon and Johannes Koell. *Wanradt-Koell'i Katekismus* [The Catechism of Wanradt and Koell]. X.2615, Tallinna Linnaarhiiv. Wittenberg, 1535.

Index

Y

Z

CPSIA information can be obtained
at www.ICGtesting.com
Printed in the USA
BVHW060224200222
629504BV00002B/61